Mystery of Photons and Light

By Steve Preston

2nd Edition

2018

Table of Contents

Should You Read This Book?

You look up at an orange sky and view blue trees while walking through a field of red clover. No; you are not on a distant world and this could be your world one day, but it would take a lot of work. These are examples of light. This book is about photons, perception, and something that is not well known-we sometimes call Light. Taken for granted, most have no idea what it is and why it is. Even our greatest scientists misname light as in the "speed of light" which has to do more with photons and almost nothing to do with light. Besides trying to determine what light is, we will also discuss how light effects life and even look at the strange descriptions of light in the Bible are only now beginning to be understandable.

Moses Issue

Let me start by looking in the first book of the Bible and its very first verse. Here we find that during the 1st Age, the Creator God made "Light". Strangely, this was done well before the Sun was made during the 4th Age. Didn't Moses, the writer of the book, know that we receive light as photonic emissions from the Sun and the reflections off objects we are trying to view? Was Moses trying to confuse people? Is it a mistake? Is there light without a light source? If you have been confused about this, I think the book can help.

Do I See What Others See?

If you know something about photons and electro-magnetics, you know that a photonic emission with a vibrational component or wavelength at 630nm is perceived as red. Have you ever wondered if what you know as red looks the same for other people? After all, the processing of photonic energy by the brain cannot be completely the same for everyone. Do some see what I think of as orange or some other thing? If this is a question you have had, this book can help. By the way, please understand that radio waves and light are identical. If someone transmits two slightly different radio signals that are separated in frequency by only a very tiny bit, it could start glowing "red" and the only reason we don't see what is normally coming out of a radio transmitter is we block those wavelengths as being too disturbing for our brains.

What You See

Have you ever heard the term, "What you see is what you get?" It is the exact opposite. What they are trying to say is beware of what you don't see. In this book I'll step you through what you don't see. It won't be always easy, fun, or even understandable, but after a while it will help open your eyes.

The Blind

Blind people receive the same photonic emissions others see, but they cannot process them and, generally, stay in a darkened environment. When they dream, light is opened to them and they can see faces, images, and other things that their eyes have difficulty with. Are they seeing light in their dreams?

One blind man told me that he had the sensation of color from sounds. Now that is going to get us really crazy as sounds are pressure packets that cause particles to vibrate more quickly and less quickly at a certain rate. This variation of vibration sets up gravitational pulses rather than magnetic pulses associated with what we typically call light, so let's ignore this strangeness for a little bit so we don't get confused.

X-Ray Vision

Just imagine for a minute that you could see all photon wavelengths from the visible to x-ray. The first thing you would notice is that people don't end. They continue beyond their skin as photonic energy in the form of heat is emanated out of the body. As you look at the person, you would not really see them clearly. They are simply shadows as objects behind them are in view as well. Anything that is hot nearby, possibly a lamp, would begin to take over what you see. Colors are all dimmed as longer wavelengths turn everything more reddish. You would very quickly want to limit what light you could see so that you could see. If you ever wanted x-ray vision, you need to read this book and you should change your mind.

Do Photons Turn to Light?

Look up at the sky. Light is everywhere but no one, I mean NO ONE, knows what light is. I know it's <u>something</u> that <u>helps you see</u> things as it reflects off them, but the harder question will be; how do photons turn into light? Some people tell you photons are particles that have no mass, others say photons are things we call electromagnetic waves like a radio signal. Others get more scientific on you, but the result is the same. They tell you some of the characteristics of photons and leave most of the details about light alone because they are simply too scary. Have you ever heard, "photons are sometimes a particle and sometimes a wave"? Who in the world would believe that type of definition? To make it worse---

You don't see photons.

You simply record electromagnetic frequencies and intensities. Something mysterious happens if the desired frequencies are received. From nowhere, you <u>perceive</u> "light". You "see" changes in electro-magnetic frequency--- and someone called it perception. Keep the same frequencies but change the receivers in the eye that convert electromagnetic waves to different electrical pulses and again there may be <u>no perception of light at all</u>. This is actually great news for us as we would have a very difficult time understanding our environment if we saw radio waves. Like the x-ray vision, everything would be horribly confused. You would be afraid to open your eyes [If those were the receptors of these extended wavelengths].

Do You Wonder About Bio-Photons?

Have you heard about the possibility of curing cancer by sending affected cells "light messages" and that each of our cells cannot only transmit photons, but also, they can receive and interpret <u>Bio-Photons</u>. So much work is being done in this area it's not even funny and for us; it is making it even more difficult to define what photons are. Besides proving that halos were real and giving us substantially more insight into what a DNA is, this new characterization of "sight" will be reviewed in this book.

Let me ask you a question. Do you think it is odd that the speed of Photons [mischaracterized as the speed of light] seems to control your relative time-base? What I mean is that if you traveled at a speed close to that of a photon, time would slow down or even stop. Let me put it a different way. If everything around you was going the speed of photons and you stopped, you would instantly get extremely old, die, and turn to dust right in front of everyone that is moving.

Let me give you a more practical example.

You are moving around the earth at 1,000 MPH, as the Earth moves around the sun at 66,000 MPH, as the solar system wobbles in the Milky Way at 43,000 MPH, as the Milky Way rotates about 485,000 MPH and it zooms through the Universe at somewhere around 1,300,000 MPH. The aggregate speed is something like 16% of the speed of light. As relativity is a velocity squared function. If someone is watching all our travels from a stationary point with the major direction of our travel away from him, he would be watching people age about 4% slower than we think we are aging and we would feel nothing. If we were going towards a viewer, our sun would not be yellow, it would be red. Let me throw out some basic questions.

- *Why are low frequency electro-magnetic waves useful as radio waves and high frequency cosmic rays just kill people?*

- *What in the world does time have to do with light in the first place?*

- *Why is there such a similarity between gravity and magnetism? Is there something they didn't teach us in school?*

- *Does the above statement have something to do with light sometimes being a gravity-based particle and sometimes a magnetically based wave?*

- *How are scientists now slowing down light at temperatures near absolute zero? What does that mean to us?*

- *How are scientists now able to turn metal invisible by subjecting it to high frequency electro-magnetic waves? Does that mean if we stand too close to a microwave, people can see through us? Maybe it's a good thing the Klystron emitter is incased in the oven housing.*

- *Why are people going around and saying photons are light when almost none of the photonic/electro-magnetic wavelengths contribute anything to adding light into our environment?*

- *WHY are time and photons tied together when an actual <u>photon must go much faster than the speed of "light"</u> or it would not exist in the first place as shown below?* Notice as the Amplitude or intiensity of the photonic emission is increased the particle must travel much farther so the particle is going MUCH faster than the speed of light.

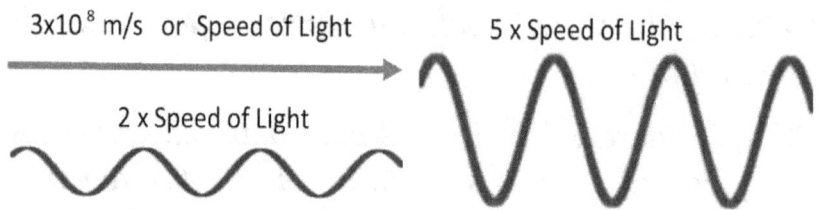

3x10^8 m/s or Speed of Light

5 x Speed of Light

2 x Speed of Light

If I can help you with just these few questions, the book will have done well and you will have begun a journey into a more enlightened understanding of this impossible mystery we call light.

The Answers Before the Explanation

Let me save you the trouble of going through some pretty odd and sometimes complicated discussions about what light is and how it is established in our universe. Here are just a few of the descriptions of light that will be defined, discussed and explained and proven in the texts following.

1. Light and matter are the same thing except time is reversed. I will have to explain.

2. Photons go faster than light. If they go too fast, they turn into what we perceive as matter. Totally different than the first statement, this will also take some explanation.

3. *Others tell you light is made if an electron is excited* and moves away from an atomic nucleus. As it comes towards the nucleus, photons are emitted and somehow this electron motion "is" light. Nothingness makes light so why is it important?

4. The old definitions *"photons are sometimes a particle and sometimes waves"* can actually be explained.

5. Understanding Light is changing how we treat human illness.

6. Light changes it characteristics the faster it moves away from you. Called "Red shift" it helps us understand all dimensional elements of our universe.

7. We MUST have a linked universe outside our own or there could be no light.

8. If two identical light-beams are emitted but one is out of phase with the first, there will only be darkness. This is similar to how noise cancellation working in headphones and how matter can become invisible.

9. According to Participatory Anthropics, there is no light without an observer.

10. A person spinning around at the speed of light may become light, so don't spin that quickly.

11. Without people there is no such thing as light.

12. It doesn't matter how our brains process electro-magnetic/photonic waves, RED is perceived identically by all viewers. While this seems impossible, it will make sense later.

13. A majority of the photonic emissions from your cells are in the ultraviolet region, but sometimes there can be visible light emitted as well. If our visual sensors could see a wider range of wavelengths, you would glow from the ultraviolet emitted and you would glow from the photonic heatwaves generated. You simply would not be able to recognize people nearly as well so it is better to have our limitation.

14. One can say light builds animals, plants and people. By pulse coded optical messages. These messages, somehow established by the DNA whenever stress is sensed, initiate organized response to outside interference and help cells all work together. Cutting a hand would initiate such optical communications to repair and replace bad elements of the organized DNA regulated structure.

15. While plants do use pheromones to communicate, they also use photometric communication similar to animals and ultraviolet light emissions can be seen whenever insecticides are sprayed. It is like photonic

emission shouts out the fear and hurt. Possibly this also aids in repair as well.

16.   One reason there are such varied results from Bio-photonic research is something called Anthropics as the will of an individual can completely change the characteristics of bio-photonic actions.   While this same dilemma, has plagued scientists for many years, we are just now starting to understand how positive responses are greatly enhanced by positive thinking, and knowing a reaction WILL occur. In the early days this was called faith. Today it is the key to great advances in understanding photons for Biophotonics and for modifying our entire reality.

17.   Somehow a dead and alive DNA are exactly the same except for one critical element. Dead DNA do not emit photons for some unknown reason. If we understand Biophotonics we may understand death and the mysterious soul better.

18.   The Bible ties light and life together and anthropic science tells us the same thing.

19.   The concept of light and faith have a close similarity, we will have to look at.

20.   The Anthropic science description of "a Cognizant Observer making reality" is not talking about a dog viewer. It is the cognition of the soul portion of a human. The portion that can control reality can also make light.

To find out what <u>LIGHT</u> is, we will have to look at relativity, a little quantum mechanics, the newer science of Participatory Anthropics, the Bible, and even how we define the universe itself. To give you a preview of Anthropics, someone asked Einstein, *"If a tree falls in a forest and no one is around does it make a sound?"*--- His Anthropic answer was, *"There was no tree."* That may seem like no answer at all, but it will make more sense as we look at how light is revealed. I know this all sounds confusing right now, but I promise you it will allow you to see more clearly [so to speak]. Before we get to old Einstein, let's just start with a few components of Photonic energy.

Photons Vibrate

Someone told you that different colors of lights were different vibrational frequencies of light and you believed them. Have you ever wondered about these innocent vibrations? I know you have seen how light splits apart as diffraction turns light in different directions as shown next.

The reason for the separation is the different vibration frequencies of various colors carried in white light. Forget the confusion about Light and Photons right now and let's concentrate on our perception of photonic wavelengths. In fact, I will call everything light until we get into more detail about the uniqueness of what everyone SEES later.

To gain insight on <u>light</u> vibrating we must look at some things we "loosely" call atoms. Now that we know atoms cannot exist in a 3-dimensional world, we have expanded the concept of the universe to be 10 or more dimensions. Some of these dimensions we will discuss to allow for a more practical definition of photons, but right now just understand that 3-dimensional descriptions have major issues when trying to define ANYTHING.

Relativity

You may be wondering what I'm talking about, but let's take relativity for a second. If you take a box traveling close to the speed of light; it no longer is a box as it is almost infinitely long in the direction of travel and tiny in the direction perpendicular to travel. If we are traveling in a circle, the box turns into a ring. To make relativity stranger; if a person is substituted for a box, the person will have almost infinite length in the direction of travel and almost no height perpendicular to that travel. To make it weird, the person will not age. If you travel in a circle, you would quickly meet yourself from the back and pass yourself up over and over making you turn into a ring. [Don't even think about the bizarreness right now.]

Time and distance are not exactly the elements the universe is made from.

Quantum Mechanics

When Quantum Mechanics came on board it got even weirder as one could destroy a particle anywhere in the universe and another would automatically appear somewhere else, completely disregarding time and space.

If you can understand quantum mechanics without Anthropics, I believe you are a great mathematician and abstract thinker, but we must go on.

Smaller than Atoms

I know you have been told all your life that atoms make up everything, but that is pretty much an incorrect thought. While the characterization of the atomic cloud does, in fact, produce things that have similar chemical structure, new science suggests that atoms themselves are made up of something that doesn't exactly exist. Einstein called it the Aether. Experimentation kept breaking down the components of atoms farther and farther until these quark things we defined as particles which combine to create electrons were found to be made up of tinier things called fermions that don't really exist. We can classify a photon as a fermion. It has no mass but acts like it does. Others include gravitons that have gravity and no mass, Gluons that somehow hold atoms together, and others. We'll look at some later, but Aether is something less than a fermion. It is what makes a fermion. I know you are wondering why we are talking about Atoms instead of photons, but remember, they told you sometimes photons made particles so we are stuck with it.

Aether and Electricity- Potential <u>but not Real</u>

Aether is the <u>potential for making a particle</u> and, in some ways, it is identical to electricity. In the big scheme of things both Electricity and Aether are only "potentials" of something. To make you mad both "must" exist. I know you thought electricity was something that was real, but let's take an example. Let's charge something to 1 million volts of electricity and put it next to something else that is 1 million volts. What happens? NOTHING because electricity doesn't exist. One must change the environment to allow it to come into existence as an electromagnetic wave or current. If there is no vibration and 2 different "levels" of electricity come in contact, nothing much happens. They will neutralize, but when vibration or oscillation is added, things become substantially different. The union produces ELECTORO-MAGNETIC WAVES. Aether does the same thing. If different Aether levels come in contact while vibrating, together they produce what we call MATTER.

What is Light/Photonic Energy?

It is this principle that will help us understand what light is, or at least the physical characterization of it. It starts off as electricity and becomes a reality with the addition of vibration. I'm not talking about the normal vibration you can readily see because this vibration is vibration of NOTHING. The faster it vibrates the more powerful it becomes. As a radio wave it brings us pleasure, as visible light it "sort of" brings life into our world. At faster vibration levels it become X-rays to see through us and faster still gamma rays to destroy us. If it vibrates faster still it can become a solid magnetic monopole. I won't be getting into that, but I brought it up for the example with particles.

What is Matter and how is it Similar to Light?

Aether is pretty much the same as Electricity. As Aether vibrates it produces something called Gravity. Vibrating slowly can produce Hydrogen and faster makes heavier particles. If Aether vibrates too fast it will become solid gravity or what we call a Black Hole. Some believe that Mass is a continuation of Light. It seems electro-optical vibrations are all lower frequencies than even the lowest level of sub-particles. A theory that combined the 2 was born.

Light/Mass Combination Theory

Now that you have been reading, "What do we know about LIGHT?" If you said almost NOTHING! You would still be almost completely correct. We are still talking about photons for the most part, but they have a place in light so let's continue.

Certainly, Photons are particles or they couldn't exist. Right?

No, no, no my friend [I see some of you got it!]. Light sometimes is sort of like a particle and sometimes it acts like there is no mass whatsoever, but there is an electromagnetic wave or frequency associated with the color of the light. [Whatever color is!] We know that the faster the "photon of light" vibrates, the more powerful it becomes. Soon, the fast vibrating photon thing becomes dangerous to humans and if its vibrating slows down too much it changes into something we call radio waves. I know you are thinking that these radio waves must not exist because they don't produce light and they have no mass but let me assure you that sometimes these photon things do act like normal matter. If you look at the following diagram, there is a wiggly line. The faster wiggling represents a prime particle vibrating faster and faster. Radio waves turn into light that turns into the deadly gamma rays.

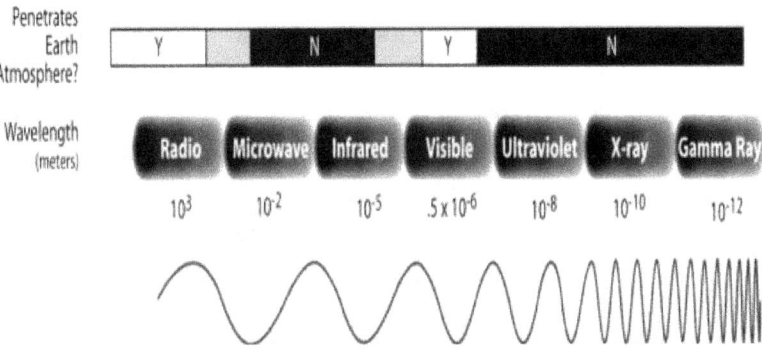

Vibrating Faster makes Matter

Like I told you every particle is actually a grouping of common vibrations. Sure enough, today the chart continues farther <u>according to some</u>. Faster than gamma this "wave thing" we have known about for years MAY turn into matter as shown below.

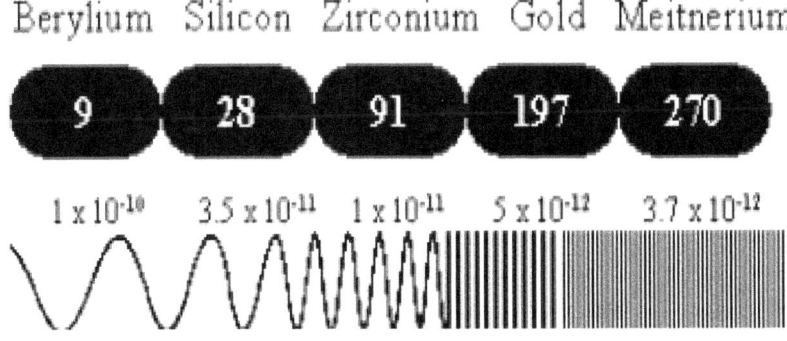

Notice that the higher the frequency, the larger the "Apparent" atomic cloud becomes. The vibrational frequencies are in the exahertz range [exahertz means "quintillion cycles per second" by the way] so don't go out and try to whistle some gold in your pocket. It simply doesn't work. If this new possible determination turns up to be factual, it means that gold may be considered some really high frequency light.

This is all I'm saying about this theory as I think there are issues with regeneration and super-symmetry with our linked universe, but I wanted you to be able to consider it on your own.

One reason not to accept this theory easily concerns speed. If photons and light were these non-descript things that were substantially less than matter, could you slow photons down?

Slow The Speed of Photons

Making a Super-atomic Cloud

The idea of this new kind of matter was first proposed in 1924 by Albert Einstein and Satyendra Nath Bose, an Indian physicist. According to their theory, atoms crowded close enough in ultra-low temperatures would lock together to form what Bose called "a single glob of solid matter which can produce waves that behave like radio waves." This so-called Bose-Einstein condensate was not actually made until 1995 but it surely is weird.

Light, which normally travels the 240,000 miles from the Moon to Earth in less than two seconds, has been slowed to 38 miles an hour by sticking light in a cold chamber. An entirely new state of matter, first observed a number of years ago seems to be difficult for light to get through. When atoms become packed super-closely together at super-low temperatures and super-high vacuum, they act like they are a single entity and when light goes through this stuff, it is slowed 20 million times.

Remember the vibrational frequency of the photons were not slowed, what was slowed was the time it took to go through a medium. I think what they really found was a way to see what matter would look like going at different percentages of the speed of light.

Speaking of vibration of photons, there is an oddness. Photons travel too quickly. Any vibrating of something going the speed of light has a problem. What if something goes faster than the speed of light?

When Light Goes Faster Than Light

Mass and Light are the same

Modern Quantum Mechanic scientists tell us; if adjacent groups of Aether have identical vibrations and are in phase, Mass comes into existence. Get enough fermions and a boson or a quark appears. A bunch of quarks make an electron and still more can gather to create the essence of an entire atomic cloud. Many atomic clouds make a molecule, etc. etc. Each fermion in the atomic cloud travels at the speed of light around something defined as a null. There is something special about this speed, but also we can see a similar defining characteristic of an electromagnetic wave where electrons are replaced with photons, and the direction is linear rather than circular. Just think of matter as spinning energy nodes and electromagnetic waves as linear energy nodes. There is more to it than that, but for us, just recognize that matter and light are essentially the same thing or they have similar qualities.

How Electromagnetic Energy is Made

Instead of starting with Aether, we simply start with Electricity and vibrate it. As a magnetic field surrounds it, it immediately travels the speed of light. <u>That is, it has a forward motion that equals the speed of light.</u> The following diagram illustrates the point.

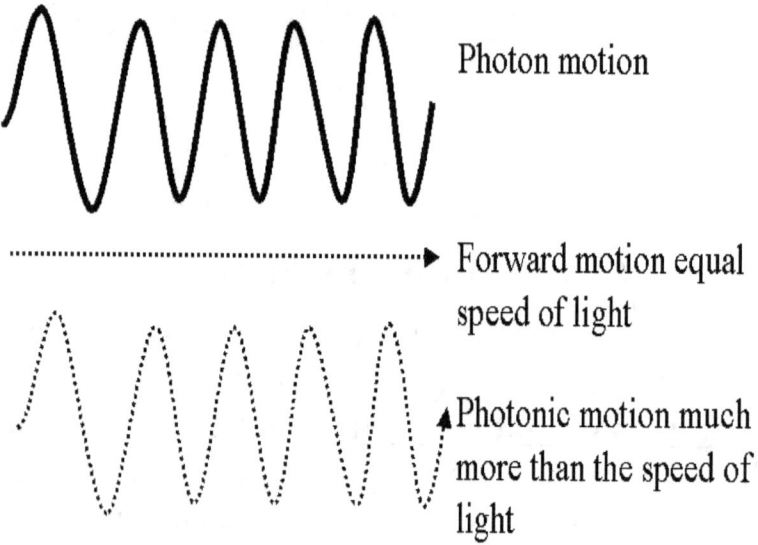

Photon motion

Forward motion equal speed of light

▲Photonic motion much more than the speed of light

Light is Faster Than Light

If the forward movement is the speed of light, the oscillations represent the travel experienced by the photon itself. That zig-zag motion assures the photon is going very, very fast. The zig-zag could be described as back and forth and it makes absolutely no difference. With Einstein telling us that mass goes to infinity at the speed of light, we determine that faster than the speed of light is like <u>going backwards in time.</u> This backward time idea disturbed me so I thought I had better figure out more about this unusual thing we call light so I can sleep at night. We'll talk about this backwards in time later as it will certainly help us define what photons and light are. Even if we don't consider backward time there is confusion we must deal with.

Light and Matter

Let me start again because I'm going to have to recondition you a little on this subject, but that does not mean what I'm telling you is wrong. While I indicated that those forcing Matter to be identical to Photons may have an issue, that does not mean photons and matter are totally separate either.

What if I told you Light and Matter were <u>almost</u> the same thing?

You would have a problem, I think, about the statement, but today, scientists classify matter as something they call out-waves and electro-magnetic [photonic] waves as something they call in-waves. If you could see either one with a reference for the other, they would look the same. Both act the same, but they also have interesting differences. Out-waves are simply vibrations [of Aether] that start in our universe and leave. In-waves are simply vibrations [of electricity] that come from a "linked" universe and cause stresses we call energy as they leave that universe and come into ours. Don't be worried if this sounds like hogwash right now and don't be worried about having a number of universes tied to ours. They don't hurt anything but without them, there can be serious ramifications. Many mathematicians tell us there must be at least 10 interconnected universes. Our Bible claims 8, and many other ancient historical records claim 12 universes. We'll stick with just one in this book as it is easier for me to wrap my head around. To get a perspective on electro-magnetics, Dr. Wolff may be our best source of information. He is important in this journey as he continued Einstein's work and brought answers that were troubling Einstein until the day he died. Before we get to Dr. Wolff, let's look at some more general concepts from Einstein.

Einstein on Light

Albert Einstein worried about light just like we do today. While the concept of photons was around, there were many issues with what they were and even what matter was. One thing he knew was what a radio was. Here is how he defined them.

What is a radio------"*You see, a wired telegraph is a kind of a very, very long cat. You pull his tail in New York and his head is meowing in Los Angeles. Do you understand this? And radio operates exactly the same way: you send signals here; they receive them there. The only difference is that there is no cat.*"

If I were to put a definition together about photons and light right now it could be similar.

A photon is like a very long guitar string. Depending on how it is plucked at one end determines the sound [or color in this instance] that is generated. <u>Light simply removes the string</u>.

Of course, that doesn't help us much, but it has a level of insight. So, we will look at Einstein and work our way to methods that allow us to, sort of, correct his observations. He and just about all of the main theorists of his day really believed that atomic structure was <u>nodal rather than physical</u>. What I mean by that is ----- Well— Let me just show you what Einstein said again.

<u>Time and space and gravitation have no separate existence from matter.</u> Physical objects are not in space, but these objects are spatially extended. In this way the concept 'empty space' loses its meaning. Since the theory of general relativity implies the <u>representation of physical reality</u> by a continuous field, the <u>concept of particles</u> or material points <u>cannot</u> play a fundamental part and can only appear as a limited region in space where the field strength / energy density are particularly high.

Matter Never Ends

OK! This is weird. He is saying <u>matter never has an ending,</u> the density simply gets less as one moves away from the center. Let's investigate a little further.

Invisible Vibration Makes Matter

Besides never ending, he implied matter doesn't exist in the normal sense. The only thing there is in existence is the universe. It's sort of "all one glob". Certainly, things are all around us so let's investigate just what these things might be. Einstein, again, helps us out. He tells us that vibrating waves of nothing make up everything. In order to give the wave or "field" character, he called the vibrating nothingness the Aether. Let's, again, read what

Einstein's own words.

"Since the field exists <u>even in a vacuum</u>, should one conceive of the field as state of a 'carrier', or should it rather be endowed with an independent existence not reducible to anything else? In other words, is there an '__aether__' which carries the field; the aether being considered in the __undulatory state__, for example, when it carries light waves? The question has a natural answer: Because one cannot dispense with the field concept, it is preferable not to introduce in addition a carrier with hypothetical properties".

There Is No Empty Space

Einstein knew that matter only existed as this Aether stuff which was some invisible vibrational or "Undulating State". Any place that had no Aether could not sustain time itself. Here again, is what he had to say.

There exists an __Aether__. According to the general theory of relativity space <u>without __Aether__ is unthinkable</u>; for in such space there not only would be no propagation of light, but also no possibility of existence for standards of space and time, <u>nor therefore any space-time intervals in the physical sense</u>.

Individual Universes

There is one more very important thing to bring up about Einstein and that is what the theory of relativity actually means. <u>It means that the observer defines "time"</u>. If someone is traveling close to the speed of light shines a light in front of him, what happens? <u>According to this theory, the light will be going the speed of light to the observer which means that one could say that it will go 2 times the speed of light</u>. Certainly, it is not that simple, but the theory has been tested and proven again and again. Another way of saying this is that each observer has his own universe, so to speak. Let's again look at Albert's own words.

"The second principle, on which the special theory of relativity rests, is the 'principle of constant <u>velocity of light</u> in vacuum.' **[To an observer]** *This principle asserts that light in vacuum always has a definite velocity of propagation* **(independent of the state of motion of the observer or of the source of the light).**

He really was talking about the speed of photons rather than light, but still interesting information.

1954 Reassessment

Dimensions attached to physical shape are dashed to bits and time starts having issues as well in Einstein's Relativistic world. By 1954, Einstein was all but beat as he tried to hold on to a universe with particle defined dimensions and some magical interaction required from people. Here is what he had to say.

According to the theory of Newton <u>the stellar universe ought to be a finite island in an infinite ocean of space</u>. This <u>conception in itself is not very satisfactory</u>. It is still less satisfactory because it leads to the result that the **<u>light emitted by the stars and also individual stars of the</u>**

stellar system are perpetually passing out into an infinite space, never to return, and without ever again coming into interaction with other objects of nature. Such a finite material universe would be destined to become gradually but systematically impoverished.

Ouch! Einstein realized that if the Universe is alone, soon we would have no light at all. Besides that, it would have not matter and----IT WOULD HAVE NO LIFE.

"How could this be?" he wondered. In his equations it clearly showed that the universe was a sphere, but if that were so, we would be losing energy every day. Depression continued after it was noticed that Dr. Hubble's red-shifts were quantized forming a pretty important observation that will simply have to wait until we get to Participatory Anthropics or it will confuse you more than you already are. Instead, let's look at how Dr. Wolff helps us.

Dr. Wolff on Light

Dr. Milo Wolff started where Einstein left off in defining what <u>light really was</u>. While Einstein had initiated the thought that matter was simply undulating nothingness [made of Aether], Dr. Wolff adds a new level of insight that will get us a little closer to understanding what light is. He gives us a defined picture of matter and light that is completely vibrationally based and uses the in-wave, out-wave description as I mentioned previously, but he explains it better than me.

Explaining the Perception of Matter

Wolff-*It is then quite simple* [I hate it when someone says that] *to show that: The discrete 'particle' effect of matter is caused by the **Wave-Center of the <u>Spherical Standing Waves</u>**. The discrete 'particle' effect of light is caused by discrete Standing Wave Interactions/ Resonant Coupling.*

OK! I'm still confused, but please understand this is a great first step. Dr. Wolff has completely separated the physical characteristics of the universe into NODES or intersections or what he calls "Wave Centers" of these wave things Einstein tried to characterize as Aether

[Undulating nothings]. By describing them as spherical, we can see that while they travel at the speed of light, they really don't go anywhere.

Just to be clear let me tell you that we are talking about Aether, the "potential" for matter, vibrating. Only vibration establishes existence; once something stops, it ceases to exist. This includes photons and light.

Where these vibrations cross paths there are areas of "no apparent vibration". Dr. Wolff called these standing waves. The standing wave initiates its own characteristic "Out Waves" which makes it appear to have mass. Think of these standing waves as the center of the nucleus of an atomic cloud or think of the as Aether. I hope that clears it up for you. If not, I'll redirect an answer later. Let's continue. Einstein worried about this a lot because all the energy of the universe was leaving. He had no in-waves in his theory.

Wolff Defines Time

Wolff-Time is caused by Wave Motion, as spherical wave motions of Space which cause matter's activity and the phenomena of time.

While this sounds like gibberish, it does something for us. It eliminates physical attributes. Hopefully you are not still thinking the universe can be defined by "length, height, and depth" inappropriate dimensional qualities. If you can see beyond those things that were drummed into your head over and over again, we MAY BE ABLE to define light. Right now, just sense mass as these ever-growing spheres of **<u>undulating nothingness</u>** [as Einstein would have said]. Here is Dr. Wolff again.

Forces Are Defined

Wolff- *Forces or Fields are caused by wave <u>interactions</u> <u>of the Spherical In and Out Waves</u> with other matter in the universe which change the location of the Wave-Center and which we 'see' as a 'force accelerating a particle.*

Oh boy, I think I will give you a minute before we try to define this.

-

-

-

-

-

-

-

-

-

-

-

-

-

-

-

Ok! Here is another bit of time before we push forward.

-

-

-

-

-

-

-

Let me try to establish a general concept here. Emanations from inside the universe outward are out-waves. Emanations that push these waves or react with these waves and are <u>initiated outside our universe</u> and are the in-waves Dr. Wolff described. While this part seems simple, it was the major stumbling block for Einstein as he determined that the space outside the spherical universes were completely empty of these undulations. In essence, our universe cannot exist without a linked universe giving us these in-wave things that bring electromagnetic energy into our universe as out-waves continue to leave it. As they are introduced, they affect the characteristic placement of all of the nodes which we characterize as Electro-magnetic force which can be identified as any force with some manipulation of an out-wave null.

The linked Universe Has the Same Issue.

Let me tie this together. Our linked universe out-waves leave that universe and go into our universe and our out-waves leave our universe and become In-Waves as to our linked universe. ----Are you seeing a pattern??? Our universe is stable because we have a linked universe and a linked universe cannot exist without its linked universe. They are symbiotic and feed on each other.

Quantum is Defined

Dr. Wolff uses his same omni-observed definitions to show how everything seems to go towards specific "quanta".

Wolff-Quantum Entanglement is likewise caused by the Interaction between the In and Out-Waves and all the other matter in the universe, thus matter is always subtly connected to other matter in the universe (i.e. matter is large not small, we only see the Wave-Center and have been deceived by its 'particle' effect).

Wow! Here we have a true nugget. Matter is not small, even if it is only some vibrating nothing and what we think of a matter is a deception.

I know this is pretty heavy and we still seem to be far away from defining light. What he is saying is that the ends of things aren't really where you believe them to be. They simply have fewer interactions as they emanate to greater distances and cause less and less affect. The quantum effect is actually the effect of the various vibrational waves that cross each NODE. Emanations from the node or standing wave must be characterized by the vibrational patterns. Therefore, apparent particles would appear at the various vibrational bands of out-waves.

These out-wave things, in order to exist at all, have a peculiarity we MUST discuss. They go backwards in time.

Just like everything else in our universe there is a stabilization of matter, energy, order, and time [called Super-Symmetry]. That is, forward time must be countered by reverse time or soon all-time would be lost.

Many characterize "negative time matter" as anti-matter, which makes sense, but as the normal out-waves from a linked universe enter our universe they have to become in-waves to use. The image below shows this recycling of matter, time, and electro-magnetic energy we call photons. Don't worry about the Soul/Spirit stuff right now just see how everything works like a machine.

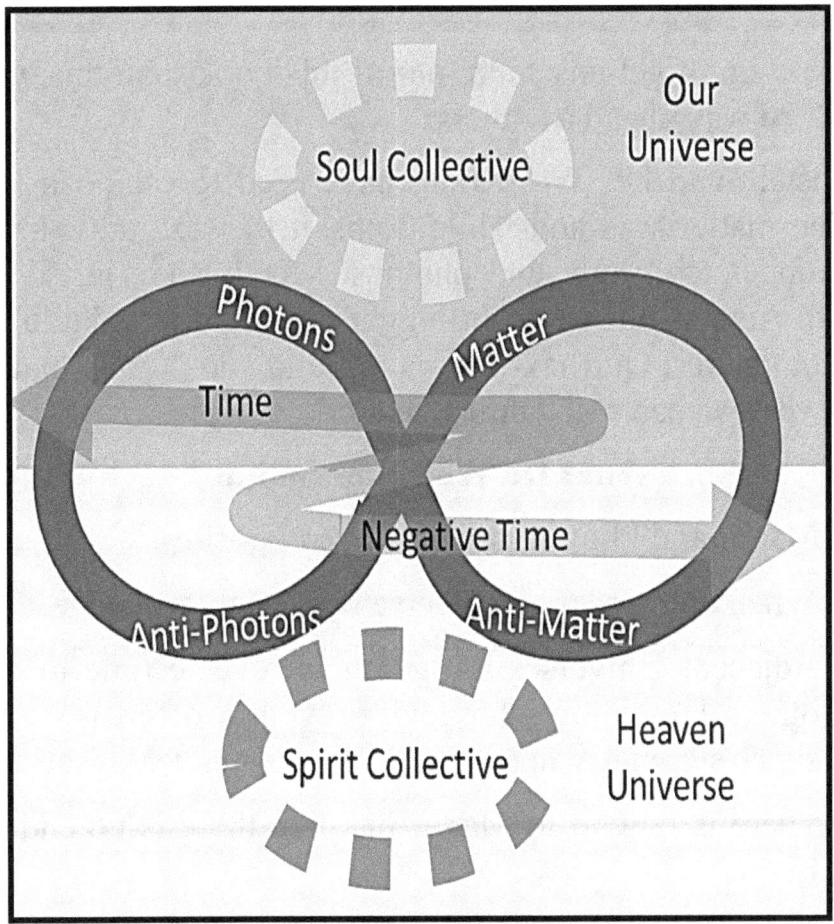

Think about this just a little. In a linked universe, matter-waves leave that universe and trickle into ours and become our in-waves. <u>It means that negative time matter is forward time electro-magnetics/photonics</u>. I'll bet you were thinking gravity from matter and magnetics from Electro-magnetics had a similarity------I'm way ahead of you. <u>They are identical, but backward in time.</u>

I already told you Aether and Electricity seemed almost the same, well-----One is backwards in time.

If you don't get any more out of this book than this, you will be way ahead of others.

Einstein and Dr. Wolff both have tried to establish just what matter was and while doing it they began helping us understand light and photons. Mostly they just dealt with Aether, but we want bigger things so let's build up some matter. Our next step is to build sub-atomic matter or what we can call "almost matter".

What Do We Know So Far?

- Matter and Photons are universe opposites.

- When one crosses a universe barrier, they change.

- Adjacent Universes are going in reverse time to the other. This allows for total recycling of energy and super-symmetry needed to make our universe work.

- There is no end to Matter just like there is no end to Light.

- This is a new characterization but seems to be appropriate here. Photons travel <u>perpendicular</u> to

Matter. Therefore, both are unaffected by the other unless a vibrational difference is noticed.

- Photons, as in-waves, make stresses and force that hold matter together and allow for work to happen in our universe.

- There is no such thing as empty space. Electricity or Aether are holding those areas together.

- Matter and Photons are both nothing. So long as they vibrate, they have "Apparent" substance.

Time and Distance Issue

This last one will be expanded on later. Time and distance have no real meaning in our universe. Of course, if time has no meaning, vibration has no meaning and this whole thing can fall apart so forget I said this for a while as we get into something less than matter.

Sub-Atomic Particles

If you have wondered about anomalous characteristics like levitation, disappearing and reappearing, and even spontaneous human combustion, they are all supported by new developments and discoveries accomplished by something called the "Tamashii Project" of subatomic particles and other similar projects and theories. Before we investigate details, let me say I'm sorry for the technical detail. It is not my intent to make equations fly around here so most of the data will be greatly generalized to make is ~~simple~~ less difficult. It is, however, important to open your mind to this concept. Without this backup, some of the details presented in this "liberation of concept" will not make sense and probably will be disregarded as too fanciful.

Sub-atomic particle research now is old science, which goes a long way in showing how levitation and most seemingly unexplainable phenomenon are possible. It also describes how photons miraculously appear, how atomic structure is determined and held, and how mass can be added and subtracted without time dilation. Everything we thought we knew has collapsed around us over the last few years and what has emerged is "easier" to understand because it doesn't have to have anomalies that could not be satisfied by the general knowledge".

Let me explain.

Old Atomic Theory

In the past, we were taught atomic theory. In that basic theory, the atom was the smallest building block and all things are made from 115 different types of atoms. Scientists started to punch holes in the theory. They found Gluons, Bosons, Gravitons, Quarks, Photons and other particles much smaller than atoms. Here is the most fascinating part to me. Sometimes these particles just disappear [just like photons]. I'm not talking about turning into Aether, I'm talking about something entirely different. All the atomic theory was in jeopardy, but the theory continued to be taught---just to mess up children's minds.

Photons

People started questioning things and wanted to look at the makeup of "light". They asked, "Where do photons actually come from in atomic theory anyway?" What I was essentially told in college was, *"A photon was sometimes a particle and sometimes an electromagnetic wave. Just believe it and don't ask questions!!"* The questions are only now starting to get answers and some of the answers make it look like levitation and even element conversion are both possible. This is neat! Maybe the ancient idea of changing lead into gold wasn't so wrong!

After all, the colleges are teaching that particles can miraculously convert themselves into electromagnetic waves. How much easier is it for a particle to simply change into another particle?

In school you may have been taught that if a photon's vibrational period is 1×10^{-6} seconds it has special properties called "Infrared". Even if you didn't take that class, the photons are infrared. I just brought up college to make you think I was smart. Of course, infrared isn't red at all. It's invisible to our eyes. It should have been called <u>Infra-invisible</u>. It was named a long time ago so we are stuck with it. The different colors of light miraculously mix together and somehow turn into white light even though you always thought that mixing colors together should make the colors darker and darker until everything was black. While you see white light every day, there is no such thing as white light. It is simply the combination of a whole bunch of colors. Something else that is strange is that Magenta [mixture of red and violet] is not produced in nature as making violet vibrate faster makes ultra-violet and making Red vibrate slower make Infrared. I'm not going to bring up Magenta anymore as it is another confusing point.

As a photon's vibrational frequency increases to 400 Tera-Hertz it suddenly becomes "visible". As it speeds up even faster, the photon becomes something called Ultraviolet [800 Tera-Hertz] and causes cancer [We will be looking at this cancer phenomenon closely as it will get us closer to the truth]. Faster still, it becomes an "X-ray" which can see through just about everything and then a "gamma ray" that can destroy tissue. All this instantaneous changing and complete modification occurred in a tiny particle and it was accomplished [without putting in huge amounts of Fusion energy associated with those Hydrogen Bombs]. I know it sounds absurd, but people accept it every day without questioning. Here is my question to you, *Just how does the photon particle change to another substance with entirely different properties?*" We never answer this "simple" question.

You would think someone would stand up and say that is a lot of malarkey, Light cannot just become dangerous, but almost no one does.

Photons and Gravity

Other new research tells us that a Photon can better be defined as nothing more than a Boson emitted from a particle-group that has absorbed energy. Because of the energy boost, it must now eliminate the energy to insure stability-and here is the most important part. The most common energy absorbed is something we call gravity. You'll be shocked to find out that gravity vibrates just like everything else. The vibration level defines the gravity.

Therefore, the typical way to make light is to make an

object appear to have less mass and, therefore, have less gravity. [I'm still way confused, so let's investigate some more.]

Don't get too bogged down in this initial stuff because it will begin to make sense soon. Besides the gravity connection here is an important part to consider from the training you got in school. "Particles" vibrate really, really fast [on the order of 60 exahertz (quintillion Hertz)] and electromagnetic wave photons vibrate much, much slower. As I mentioned before, one theory is that mass is simply very high-speed light, but that is not nearly the complete answer. We are told a "block of light" goes in and out of the particle-wave things all day long. So, what does that mean?

Two Answers

From the high-speed light theory above we can surmise one way a photon CAN be sometimes a particle and sometimes a wave is by changing the vibrational component. If the vibration changes very quickly; when it is really fast, it is a particle and when it slows slightly it becomes an electro-magnetic wave. From Dr. Wolff's theory we can surmise light going backwards in time is normal, but if time is reversed, the photons would appear to be matter. If that is completely understood, I will go on.-------- OK! I'm going on anyway and try to make it make sense later.

Atomic Fusion

Here's an oddball question; if hydrogen has one electron

and one proton and helium has 2 electrons and 2 protons, can you put 2 hydrogen atoms together and make helium? The answer has been, "You can't do it without the exchange of a substantial amount of energy associated with Atomic Fusion". It has something to do with what we learned in school that was called NUCLEAR force [The force that "allowed" atoms to stay together with large numbers of protons and electrons.]

According to the dictionary, a nuclear force (or nucleon-nucleon interaction or residual strong force) is the force between two or more nucleons. It is responsible for binding of protons and neutrons into atomic nuclei.

People create massive atom splitters to break the nuclear force and make subatomic particles of all types. As the atoms are degenerated, there is a fear that the energy created during the separation of atomic particles could cause disaster or create a fusion disaster like a nuclear bomb or other scary things. One of these cyclotrons in Europe generates so much energy that there were concerns that they <u>might create something called a black hole</u>. If you think I'm crazy, you had better stay away from the guys trying to explain just what a black hole is. If one of these black hole things happened, the earth could explode and we would all be sucked into the next universe and if any survived the trip, who knows what we would find. Let's not think about these black holes and try to stick with things that are more manageable like nuclear force.

Unfortunately or fortunately, nuclear force can be controlled. The force that holds atoms in quantized numbers of particles with quantized amounts of energy and density apparently is a vibrational force. Here is one of my interests in this whole thing. Maybe we can get around the fusion reaction requirement that is so very dangerous to our existence with vibration.

Non-Fusion Manipulation

In a vibrationally controlled world, trying to modify this "APPARENT" "nuclear fusion" is not the only way to manipulate atoms. One can affect atoms by affecting the characteristic vibrations of the component particles or by affecting the apparent vibrational element of the entire atom {particle group}. These manipulations can and have been done without the huge explosion of a nuclear bomb. In fact, they may happen as part of a natural affect which emits photons.

By modifying particular frequency components of the atom, the various groups of sub-atomic particles, apparently, can and do become invisible, or will retrograde to another state, or may lose or gain their associated particle-mass energy by emission of that elusive photon thing.

That doesn't sound like light but it more closely defines how it is produced. I know that is a big statement so I'll get you more confused by defining invisibility. You can go back to just saying if this light "thing" vibrates faster its structure changes if you want. I don't care. The thing

I want to bring up right now is that invisibility is somewhat different than you previously thought.

Invisibility Observation

Many times, invisibility has something to with another weird thing we call gravity.

All particles and groups of particles have gravity but not all particles have electromagnetism, so they may not all react to one another nor would they all produce magnetic fields. In fact, if two particles don't react to one another, <u>they are invisible to one another</u>. One way to think of this is that they are, in some way, shared between 2 universes.

Gravity Observation

Gravity of these particles may also be shared as a remnant of mass. One may believe that if gravity is sensed, by definition, mass must be where the gravity is. Some have tried to explain away gravitons, and other seemingly massless particles that have gravity, but without mass defined some way, the details fall apart. You can't shield gravity because the shield would have the same elemental particles. We are now discovering that there is another way to change elements besides changing the vibrational component.

A way to change elements is by modifying the gravity part of a particle just like I defined before.

[Sounds simple; Right?]

While no one actually has defined Gravity in any major way, I think we can define it fairly simply [Ha!] in the vibrational dominated universe. Gravity would be the cross member to the vibrational string in a vibrational membrane. I guess you know what that means! Like everything else, gravity and vibration are the same thing. If we attach particles to a vibrational string, perpendicular vibration would show how the vibration travels perpendicular to the vibrational string OR [and this is a big or] ----

I'm sorry! I'm Sorry! That whole thing just blurted out of my fingers. I'll go slower and I'll make pictures so that I won't sound like that John Keely character. If you don't know him, I introduce him later.

Cross Modulation Gravity Vibrations

Gravity can be considered the cross vibration of a Particle string. [To string theorists, a string is really a dimension associated with something. In this case Matter would be vibrating in 2 dimensions perpendicular to one another.] We could recognize this gravity effect as a cross modulation of the dimensional string that makes particles as shown below. I drew it circular as Particle dimensions don't move, they sort of go around in a circle. Because gravity vibrations are perpendicular to vibrations of particles they are invisible to each other. As we look at electro-magnetic waves, the model would be a straight line [Electricity] with magnetism vibrating perpendicular to Electricity [see second image].

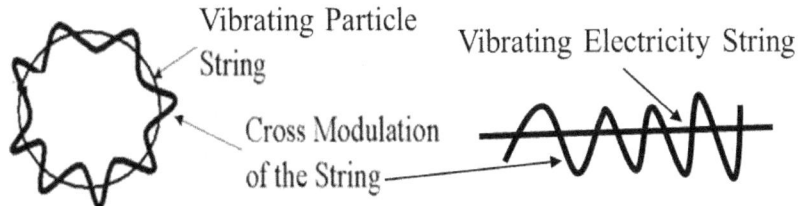

Vibrating Particle String

Vibrating Electricity String

Cross Modulation of the String

Graviton Invisibility

A graviton is a particle that has gravity and no mass. Scientists see the effects all the time, but really defining what a graviton is can be a mystery. One way to look at the graviton is 2 fermions [Quasi-mass] things vibrating at the same frequency and in opposite phase. Everyone knows what happens. It's like noise canceling in those BOSE headphones. Unwanted sound becomes invisible. In this case the particle becomes invisible even though there are 2 of them just sitting there.

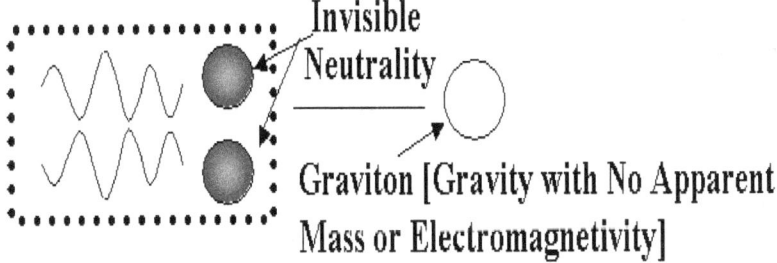

Invisible Neutrality

Graviton [Gravity with No Apparent Mass or Electromagnetivity]

Noise Cancellation

Let's take a nice set of BOSE noise canceling headphones. These things work by bringing in background sound, inverting the sound and pushing the normal sound and the inverted sound into your ear 180 degrees out of phase just like the drawing above is showing 2 "quasi-particles" vibrating 180 degrees out of phase to one another. While you would think there would be MORE SOUND for your ear to deal with, the background sound disappears just like the mass in a graviton.

Magnetism with no Apparent Photon

In electro-magnetics, the electrical field component reacts in a perpendicular direction to the magnetic field component. This effect in space-time can be similar to the perpendicular gravity field that may be compared to something we should call mass. Everything is vibrating. Gravity, Mass, Electricity, Magnetism, light and even the so-called nuclear forces are all just variations of vibrational "stresses" to the elusive single particle that makes up everything and that brings us to Einstein again.

Einstein was the premiere relativistic thinker of the present day. While most of his discoveries dealt with functional groups of particles that made up elements and how these elements reacted to each other either in this timeframe or as it was modified as the dimension of time or its reference was modified. Let's get back to how photons are traveling faster than the speed of light.

A New Theory

One of the reasons I brought up light going faster than the speed of light was because you have always been told the following: *as any object approaches the speed of light its mass grows to infinity. It's one of those can't be done things, but mathematically one can show the growth of mass.*

Photons "going the speed of light" have <u>no mass</u> so there is something wrong with Einstein's predictions ***unless*** the out of phase vibrational patterns neutralize the effective vibration and <u>the mass disappears </u> *or something else.* Let me provide you with another possibility and potential answer to going FASTER than the speed of light.

Photonic Dilemma

Let's look closer into this seemingly impossible faster than the speed of light of the photon. Some people simply say the statement has no meaning as <u>it is the effect to an observer not the traveler.</u>

If we are to try to make some sense of this we need to look at mass a little closer. Let's take that photon and look closely again. The photon itself is traveling forward at the speed of light, but the photon has an electromagnetic component which vibrates the thing. If the forward speed is the speed of light and the photon is vibrating, the actual "particle" MUST be going faster than the speed of light or the other operations could not be accomplished. Going faster than the speed of light forces an object backward in time which is where photons want to be in the first place. To make this weirder, all the different colors of light vibrate differently so they are going backwards in time at different rates. To make this oddness even more odd, it is going backwards in time **while** it is going forwards in time to allow us to see and use it. It is as if <u>time has no meaning to it</u>. If we expand this concept a little further, we see that –

Anything going the speed of light loses time relationships.

Do you remember the old concept of going into space in a spaceship at the speed of light and coming home only to find out that everyone is hundreds of years older than you? To everyone including yourself, you jumped ahead in time. Using this process, you would go backward in time if you had gone slightly faster than the speed of light. I don't recommend it as you would cease to exist, but Photons might be different.

Light is Sometimes Invisible

Like you were taught, light sometimes doesn't have mass. It disappears, but the wave of the photon carries its properties thru until it reemerges as a particle only to disappear once again.

Faster Than Light

Let's back up and see what we have. If photons are really matter, they are made up of Aether which is non-particles vibrating. Because it is a non-particle, light can go faster than the speed of light without the unfortunate effect of getting infinite mass and causing all types of problems.

How?

The way this seems to be possible is that the vibration of a non-particle is different than vibration of something we can see touch or feel. Vibrational packets seem to be areas where a photon enter into this existence going forward in time. At this high speed, the photons are not stable and crash into a different universe until they again slow down to enter in this world going backwards in time. I know you only hear that light is sometimes a wave and sometimes a particle, but I'm telling you-

> *Photons cross the Universal Boundaries-----Light sometimes is a particle and sometimes is <u>nothing at all</u>. It only has electromagnetic properties while it is in this world with an effective negative time-base.*

Lots of Energy

Why is there so much energy in a photon? Photons do all types of work and they are tinier that the tiniest thing you could imagine. Luckily for us, or as planned by our maker, this crossover requires the mass to go to zero which cannot happen without converting all the mass into energy.

Bam!

Then when the crossover back into existence occurs we have the same thing and energy is quickly absorbed to produce this mini-event horizon along the path of the photon. One would think this is some type of black hole, but the secret is in the equalization of mass before and after the transition. Our universe cannot sense this crossover. All that energy shows up whenever this marvelous thing we call a photon goes the speed of light while exiting in <u>this world.</u>

I forgot to mention that the 2 linked universes that allow for our existence share time in an unusual way all forward time direction is backward time in the adjacent world. Therefore, the idea that these photons are going backwards in time can be mathematically confirmed. Don't worry about this idea right now if it makes you uncomfortable because we will go over it some more later.

If we make a diagram of our life, it might look something like that below. During this lifetime, there was a short time going near the speed of light so a lot of living happened in a short period of time.

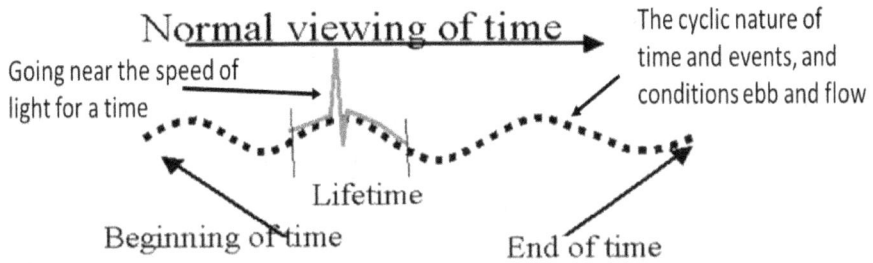

Once we have this model we need to look into something that is going to feel strange. This next concept is called Lateral Time. From a lateral-time-viewpoint, <u>time has no meaning</u> so forward and backward time of these photons can be more easily understood.

Lateral Time and Light

To help define light that is sort of misplaced in time, let's shift time sideways. As we review this concept we will get a better image of what light might be. We can potentially see the changes established by light if we view what I term "lateral time". A person viewing time laterally would see the <u>beginning and end of time "at the same time"</u> and he would see the <u>progression of light as a time-based phenomenon</u>. It would be difficult for us to determine what we are seeing, because we are not used to it. The image following might be something like what we would encounter. Images are of light and you can see all of time.

Therefore, I will result to drawings. Remember this is simply viewing everything differently rather than initiating a causal event and seeing a resulting opposite event millions of miles from the first event like quantum mechanics has presented, this should be easy. I put a lateral time diagram below. You can see that God can see the beginning and end of your life simultaneously as everyone's travel through life is piled on top of one another spatially. The excursions of the graph <u>represent light going forward and backward in lateral time's "equivalent of time"</u>. We could call this equivalence "mass resonance change" in our normal time perspective.

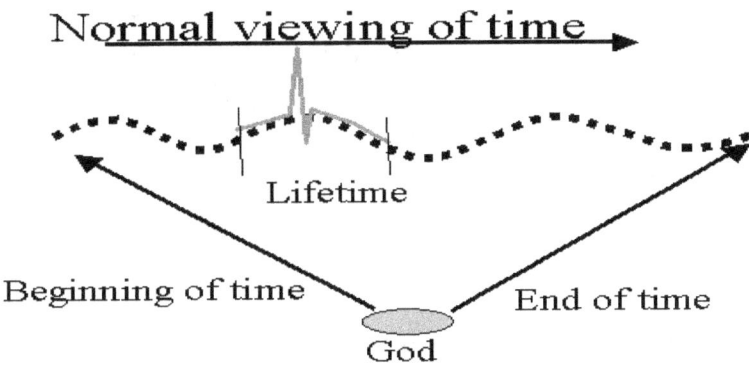

When God said he knew the beginning and the end, He would have had this lateral view of everything. He would see, for instance, your life all at once and in place of time, he would see modifications of resonance represented by the curvy line across the center of the viewing. I know this is all new to you so I will try to explain it as much as I can so you will appreciate what light really is a little better.

Now for a hard question; *if everything is really made up of vibrations only, as I have presented, what would everything look like when viewed in **lateral time**?* Hurry up; the clock is ticking. -----Come on!

> *If vibrations are emanations of modification over time, the answer would have to be a **"solid mass"**?*

This solid mass isn't a mass at all. It is simply a compressed vibration. OK! I don't know what compressed vibration of nothing really is. It is easy to write about it and make you think I know something special, but it is quite another to be able to picture what light is in your head. It all has to do with perception. As we look at lateral time in more detail we will see a stronger relationship between life and light.

> *One can say that there is an element of life that has the same characteristics as light viewed sideways. In one respect, **light is life viewed sideways**.*

Certainly, no one would suggest that a light bulb had life, but what one may find out is that there are a number of similarities between light and life and some critical differences. Before I get back into these more exotic descriptions again, let me back up a little and try to reintroduce light as defined in "normal-time". Then we will combine it more in the lateral time world and finally, I will add in the Participatory Anthropic theme of "Life Definition".

Vibrating Time

Before you can really understand time, you must first see it as one of the dimensions that make up what we call a universe. As such, it must be made up of the thing that makes up all things.

NO!

I don't mean atoms or bosons or even the almost unperceivable fermions. Time is not made of particles.

Hopefully you already guessed it. What I really mean is time is made with vibration.

Many accept the vibration of light and the other things around us, but we typically we mess up our descriptions when we view time describing it as this straight-line thing starting at the time we are born and ending when we die. We can sort of extend this same "time-line" from the beginning of the BIG BANG thing until the end of all time, but it still is in the same direction and has the same constant linear dimension. Oh!! How comfortable this description is; very visible and very easily described, but then Photons could not go backwards in time and we would have issue with a linked universe "renewing" energy continually.

Vibrating Electricity and Light

If electromagnetic fields didn't vibrate, they simply would not exist as would light itself. What I mean is that if you ever stopped a photon, you would be holding nothing. The bad part of this stopping vibration is that if we stopped vibrating, we would no longer exist as well. If we look at atoms, current studies indicate that they are simply clumps of common vibrational nodes rather than true substance. Even though everything can really be explained better vibrationally, let me first straighten out the time line so it looks more like what we perceive. As I previously touched on, if we view time from the side, we can see the beginning of time and the end of time along a line in front of our eyes. I've labeled the viewer as God, because he may be the only one that can perceive this thing. From this vantage point, everything that happens to you from the birth to death are all shown up in one instant. There is no future or past, there simply is as I showed previously.

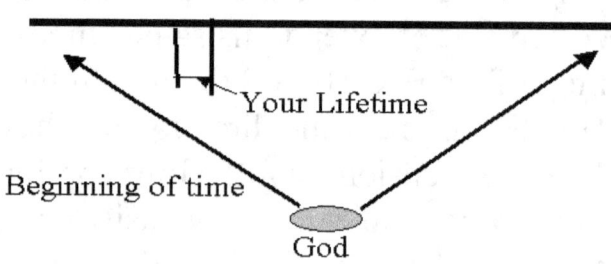

Time may be vibrational like all the rest of the dimensional strings of this universe. Instead of a straight line, on the following page, I am showing it vibrating just like everything else does. As we go through time, the hills and valleys don't mean anything to us, but the variations could be witnessed laterally. The hills and valleys might be certain cyclic pressures like destruction periods, Ice Ages, wars, and other things that mark the cyclic nature of time and God could look at all these peaks simultaneously.

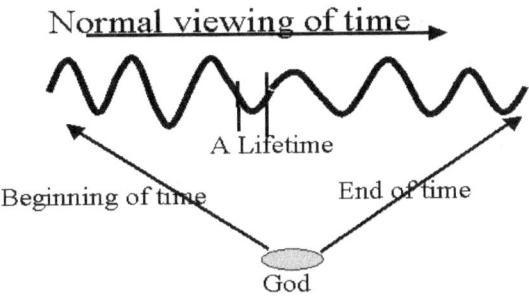

Normal viewing of time

A Lifetime

Beginning of time   End of time

God

I haven't changed time here; I simply have changed the viewpoint. Notice that a lifetime is shown as a small segment of what would be viewed. The beginning and end of time are only shown for direction. There may be NO beginning and NO end for all we know as time may be circular, so think of the wiggly line going on and on as far as you can see or in a circle as depicted in some examples of dimensional strings that explain the quantum effects noted in life.

Light Seen Laterally

While lifetimes take up a section of the time line, typically, light goes the speed of light. It is here one instant and gone the next only to be regenerated and be found again for another instant. Light doesn't actually travel along the time-line. As Einstein predicted at the speed of light there is no time. Light would view the universe LATERALLY as shown in the next graphic. Light would appear the same as life appears when viewing time normally. Normal mass would have no apparent mass equivalent and light would be represented as having this equivalent of life.

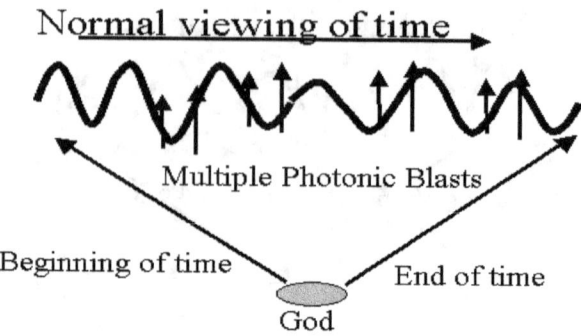

Normal viewing of time

Multiple Photonic Blasts

Beginning of time   End of time

God

Speed of Light Example

That was the easy description with no meat. Let's put on some meat and see what happens. If a person leaves here in a rocket going the speed of light and returns going the speed of light, what would the rocket look like? The answer is that the rocket would gain infinite mass along the direction of travel. It would look like a beam of light and it would not travel on the "normal timeline. The rocket and the person experience LATERAL TIME as shown next.

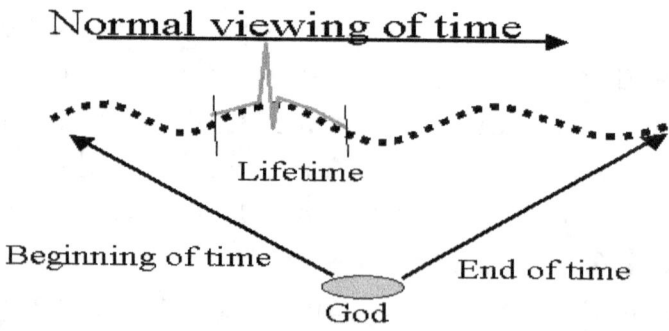

Normal viewing of time

Lifetime

Beginning of time   End of time

God

The time-line was expanded a bit to show detail the spike in the middle represents what happens when the rocket goes close to the speed of light. If a person could see what was happening outside, <u>he would see everyone's</u>

life passing in an instant. The downward portion of the spike is his return to normal home at close to the speed of light. If you haven't seen it from the earlier examples, let me tell you that you will turn into--- "light"---- if you go the speed of light.

Turning Into Light

Let's look deeper. If you could see someone who was viewing you in lateral time, how old would they get as you aged? Of course, they would not age a day because they could see your entire life as an instant. If you go the speed of light, how old do you get with respect to those not going with you? The answer is that you would not age.

A more defining answer is that if you could possibly go the speed of light you become light and are traveling in lateral time. However, all the particles in your body are vibrating so the particles making up your body are going backwards in time or they must stop vibrating.

Forget I said the backward time thing and let's move on. If particles stop vibrating, they do not exist. While this would be bad for 'LIVING" people. Light may have no reason to exist or not exist. Therefore, crossing over to adjacent universes should be possible for light. Let's look at the diagram that was presented before.

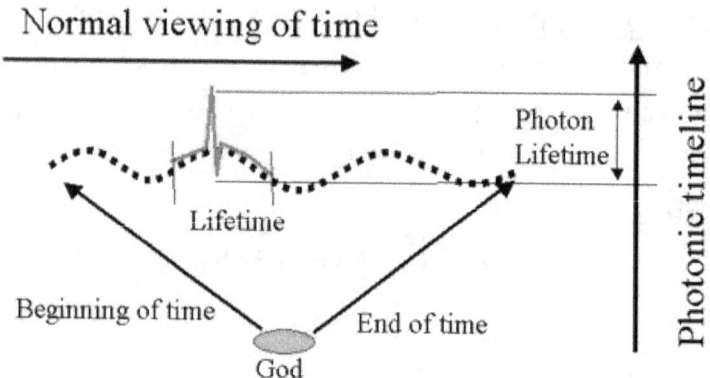

In the diagram I showed that light really had no normal time component but would look like a spire out from the normal timeline, in fact, if we were to recognize the lifetime of a photon, it more correctly is determined as shown above. Beginning and ending in the blink of an eye, but it could last for the equivalent of centuries in what I'm calling lateral time.

Speed of Light Example

In a lateral time world, what would a rocket look like? The answer would be that rocket would look as long as the entire trip the rocket took in time [It sort of converted to length]. I brought that up for a reason. What would a rocket going the speed of light look like in "normal" time? You guessed it; it would be as long as the entire trip to a viewer that was stationary.

Vibration

If vibrations are emanations of modification over time, then the rocket could be viewed in lateral time as a solid mass. I know that sounds like a black hole and matter must be being sucked into oblivion, but it's not.

Remember, we are looking at a different dimension. To the viewer of lateral Time, all the vibrational motion would now be produced simultaneously. As a particle moved over a distance, it would look like a line drawn across many more lines. Let's explore this just a minute. How do we perceive the vibrational patterns associated with a mass? The answer is that we perceive them as a mass. We cannot see the vibrations and, frankly, we cannot even understand them as vibrations. It is as if our "impressions" of things are associated with viewing matter in lateral time rather than linear time that we experience.

Photonic light would be seen as solid beams like wire and here is another thing to think about. How long would the light beam be visible? ---Da-dah-dah-da-da-dah-dah [I'm humming the Jeopardy tune.]-----"What is forever?" is the correct question.-----

I mean the image would be there forever in lateral time.

Lateral Time is the Speed of Light

I know we use this "speed of light" thing every day, but now we can define it differently and possibly more accurately.

The speed of light is when vibrations become solid. It is lateral time.

I know you were expecting that absolute zero or minus 473 degrees would make all vibrations stop, but we actually had to go in the other direction.

71

The speed of light also is the transition from linear time to lateral time.

Once we cross that boundary, time travel, by this description should be possible in forward and reverse by simply injecting yourself into the lateral picture presented in lateral time.

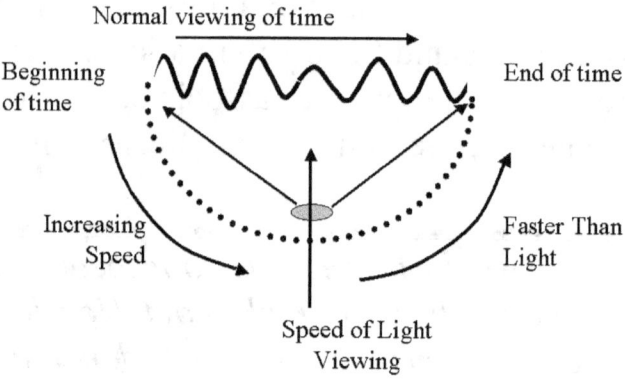

The pervious drawing, hopefully, will sort of explain what I am saying here. The curved arrow on the left shows what happens as you move closer and closer to the speed of light. What you would see is that everything would speed up for you. As you approached the speed of light things would be skittering around faster and faster. Soon there would be a blur and everything would be impossible to understand because <u>you would be seeing all time in an instant</u>.

Beyond the Speed of Light

The blur begins to slow down slowly as you go faster than the speed of light as described by the curved arrow to the right of the drawing. What you notice is that <u>things</u>

are going backwards now. Soon you can get to a level that is similar to where you started with respect to the motions of everyone, except that everything is backwards. While in this state, one could gain knowledge of the future that would now be represented as the past. Once the information is retrieved, slowing down to a stop brings you back again to "normal Time" and you have the secret of the future with you. If you noticed, as soon as you go faster than what we call the speed of light, you don't go faster. Instead, you go backwards and that brings us to an important description of time which is certainly needed for any discussion of light and life.

Backward Time

Let me tell you something else about this backward in time vision. You would be seeing an adjacent universe rather than our own. I'm not even sure you would recognize any of the elements of our universe very easily at all. For us, we only need to understand that there is something we can call a Constance of Time.

Constance of Time

Just like all things, light, matter, and life, time CANNOT simply go into infinity and be lost.

If time was only one way, soon we would have no time. Therefore, there is an underlying negative to time. We can perceive this negative time as an adjacent universe [some call heaven] where time is completely backwards. If we view this NEW world, the in-waves there would be the out-waves of our universe, their electro-magnetics would be our matter, etc. There is a strange linkage between our universes and both are needed for either to survive.

What Do We Learn From Lateral Time?

As we search for a definition of Photons and Light what have we found?

- Light going backwards and forwards in time can be reconciled in Lateral Time.

- The beginning and end of time being together allows us to witness light as a straight function of time.

- Light appears to be Matter viewed by lateral time.

- The speed of light is when vibrations become solid. It is lateral time.

- The speed of light also is the transition from linear time to lateral time.

Almost the Speed of Light

Let's slow down to just under the speed of light and look at what Einstein had to say again. Einstein told us and we have later confirmed that as we go close to the speed of light, our mass reduces and time almost stops for us. Let's say two people are together and the same age. One goes traveling at close to the speed of light for 20 years. Neither shaves until the traveler returns. When he returns, the fast guy has not aged a day and the other guy is 20 years older and has a 20-year beard, as shown below. It didn't matter where the traveler went and how he came back so long as he did it close to the speed of light. He could even stop every once in a while, to see stuff. For the guy going fast, time almost stops, just like the suspended animation "sleep" method, but this time we simply used what Einstein referred to as "relativity" which forces the speed of light to be constant to all observers. If you are going the speed of light, you must experience light "generated by you" going the speed of light, so you must not experience the time during your travels AT ALL.

Of course, they have proved this anomaly with a number of experiments, but again, this only allows the forward dilation of time. If you went on your journey for a thousand years, everyone you knew would have gone and you could never get back to them. That would be a bummer. I know this is a way too brief description of this weird phenomenon, but the fact is that this one-way relativistic time travel is not what this book is about either.

Notice I showed that the guy going close to the speed of light does not have to travel in a straight line. Let's see where that takes us.

Instead of going somewhere, the fast guy simply turns in a circle [After all that is the way the Aether travels the speed of light in matter]. It doesn't matter, the same thing happens, in the blink of the fast guy's eye, the slower guy turns old and feeble.

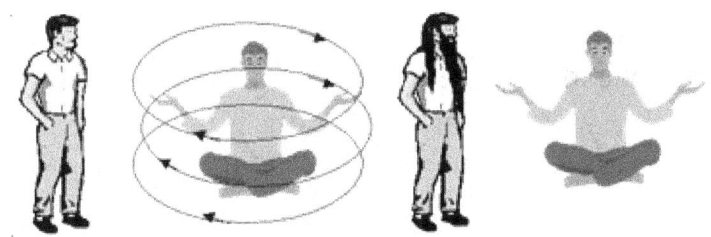

Let's look farther. Let's say the spinning guy turns on a flashlight aiming outward. What would the slow guy see? Each time the flashlight came around in his direction he would be blasted in the face with light. Because of the spinner's speed, that would be all the time.

To the slow man, he would, sort of, see his friend turn into light.

Now assume that the spinning man was not just moving around in a circle but was spinning in all directions and you see that he becomes a speck of light. The next time you see light just think of it as a spinning man and time travel may become a little closer. Once he got to the speed of light, he would be going backward in time and his out-waves would now be considered in-waves.

Tiny

I forgot one more thing in the example that sounds problematic. As the fast man spins faster he gets smaller and heavier. Close to the speed of light, the spinning man is very small and very heavy. Mass tends towards infinity as the object approaches the speed of light. While the spinning man has completely vanished except for the light shining in all directions as he steps past the speed of light. On a linked universe, he is still matter, but with so much gravitational pull he is considered a black hole or portal to the reverse time universe.

Vibrate

Do one more thing for me. Think about what would happen if the man didn't spin around in a circle, but instead, he stayed still and his body vibrated close to the speed of light. He would again stop aging and time would almost stop for him. He would decrease in size and increase in mass---all while he was standing in front of the other man. The vibrator had actually traveled into the adjacent universe, was viewing backwards time as some type of black hole thing.

This has given us more information to define photons and light with, but we still need more. Let me give you another example I think will help clear up something. In this case we will have an object move away from you and we will look at something called the RED SHIFT that violates reality.

Violation of Reality

While I'm mostly talking about light, everything must be considered in this example as it affects our reality. That is because vibration can only be defined in a set time. While length, height, and width defined in a rectangular coordinated world could just sit there, a vibrating mass has a serious problem. Today we know that things change as vibrational frequencies change. Let's look at some at the stars.

Red Shift and Gold

Dr. Hubble discovered that if objects moved away from us very fast, the colors that emanate from them shift towards a more reddish color and these "RED SHIFTS" are packetized as if blocks of things are all going the same speed from us. To the object moving away the colors didn't change, but when we watch them the colors are all different. The light changed into something else because the number of vibrating cycles had to stay the same as shown in the following graphic had to get slower for us to see it. I think the next example will show you what Dr. Hubble really found.

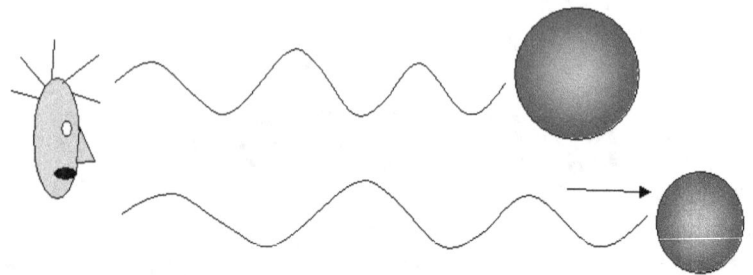

Same number of vibration cycles are created but must be stretched as the object moves away for us to see them.

Emission Spectrums

Let's think about this whole red-shift thing for a moment. Every atomic cloud emits a certain group of colors when excited. The chart below shows the different emissions of Hydrogen, Lithium and Sodium [Na]. While Hydrogen is mostly red and blue; Lithium is red, yellow, and blue; and sodium is almost completely yellow.

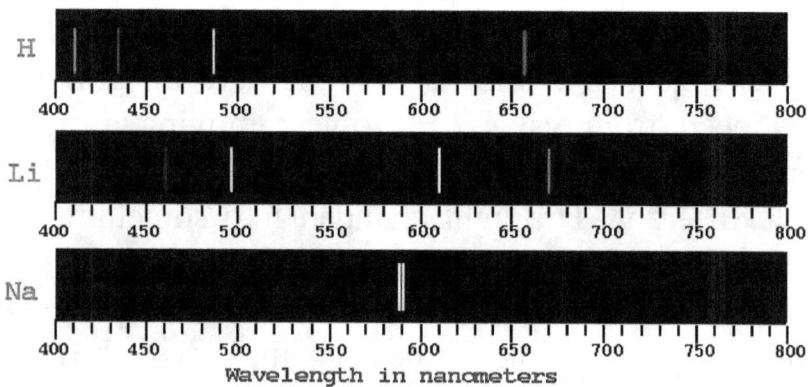

The more complex the Atomic cloud the more variety of photon emission we can expect. The Spectro-view of

Calcium, Magnesium and Neon following show a wide assortment of different types of photon generations.

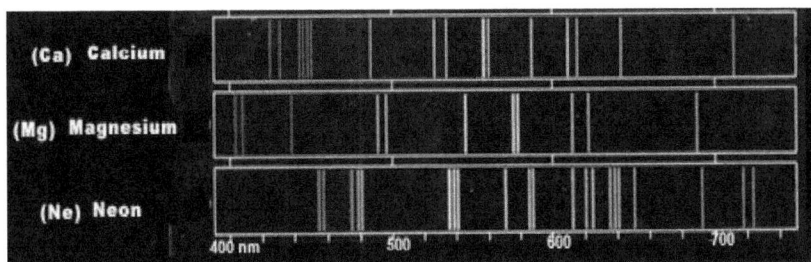

The way all this happens is that Electrons get excited and jump farther from the atomic core. When they find out they shouldn't be there, they go back to their "Normal" place, and this produces a force or photonic energy we see as light. The following diagram sort of shows you what I mean. When there is a substantial amount of energy released, the color is more bluish and the least amount is reddish.

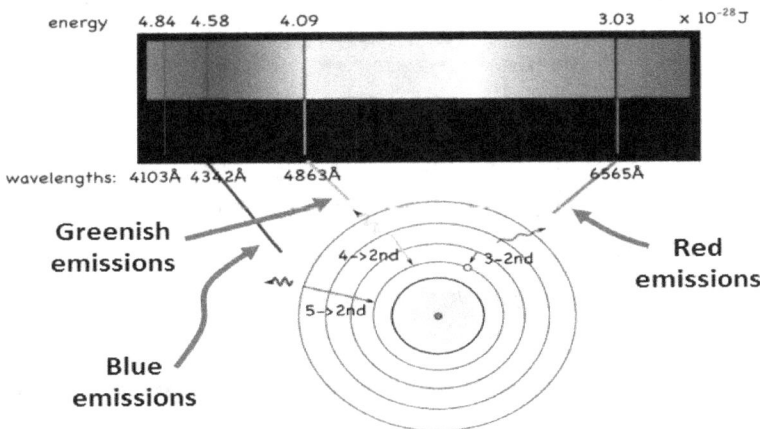

By this diagram we can understand that the atoms going away from us are getting smaller so less energy is required to get electrons back into their place [only when viewed by someone who is seeing an object [like a star] leaving at very high velocities. While this is an accepted characteristic, what it is showing is the universe associated with the object moving away from us and our entire universe stays in sync with 'OUR REALITY" by changing vibrational patterns and vibrational patterns determine matter.

Let me say this again. Time stays in sync by changing matter.

What this red-shift means to 'Vibrating" matter is that if we could see the things on the object moving away from us <u>they would **not** be the same things</u> as they would be if there was no relative motion.

Lead objects might be gold objects or we could see radio waves or any number of things that would make us go crazy. It's best for us not to look.

If the moving away guys could see us, gold objects would be lead and our bodies might be some type of bright blue material or our clear atmosphere may be completely solid. Here is the most important part of this. If a person is moving away from us at near the speed of light, he would gain mass according to Einstein, but he would reduce his mass to lower vibrational elements according to Dr. Hubble. Soon the person you were viewing would be invisible as his vibrations stretched enough to turn him into mostly oxygen rather than mostly water.

If you go the speed of light, you would turn into Aether in the reality of the viewer.

Let's look at the opposite view. If you are going fast and looking at everything around you and you see a tree. If you are going fast enough, the tree has no width at all in the direction of travel. There is no way the "line" that you see is made of wood and plant cells in your high-speed world.

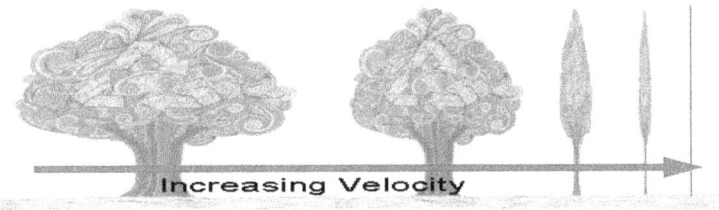

Increasing Velocity

Einstein Confirmed It

If you ever wondered what Einstein's famous formula was all about, let me tell you.

$E=MC^2$ is saying $M= E/C^2$ or Energy and mass are EXACTLY the same thing when there is a differential vibration or speed associated with the speed of light. You cannot have what we call matter at the speed of light so you would be Aether.

By the way! I am not writing Aether like "Aether bunny". I don't write with a lisp. I want to make that clear.

If we assume mass really does exist, we can characterize it in the same way as light by determining its resonant frequency or wavelength. Some believe that mass and light are actually the same thing, but I am not addressing that in this book except in a cursory way for completeness, right now I am just showing similarity between Photonic wavelengths and matter wavelengths. This is done by using the Tamashii model.

Tamashii Model

Now I think you have a basic understanding matter and light and how they change depending on speed so I'm going to quickly go through the general Tamashii model. This model expresses all elemental parts of atoms, including the quarks and bosons, as combined groups of particles which change characteristics depending on their spin motion. My version certainly is much less detailed than the complete model introduced to the world by J. Newcombe Hodges some 20 years ago, but I think it will be enough data to bring in my own point of view. Even if you don't understand the invisible subatomic components yet, don't worry. It's one of those things that sort of grow on you the more times you say it. I'll say boson, or quark, or fermion, or photon from time to time in this book so you will get comfortable and happy.

Tamashii Invisibility

As I stated previously, correct quantity and frequency of Bosons makes things invisible and when I say invisible I mean containing absolutely no mass. I know you have been told every physical thing has mass, but now you know that is not correct. Light, for instance, has no mass and neither do particles called gravitons. If particles in close proximity have similar spin motion {frequency} they can react with one another. Particles with different frequencies are unseen by adjacent particles and do not necessarily react. This may also mean that the particle is completely invisible to an outside observer. Because of this elemental property even a boson-sized particle can go all the way through the earth. The earth would not register any mass with respect to it and the invisible boson would register no mass with respect to the earth. The mass of each simply would not exist to the other. Stay with me here because that is sort of what causes light as defined by these new sciences.

Just think of mass as an illusion of vibration and you'll be fine. Oh! That's right, the vision of a frequency thing is just as difficult to imagine. Especially when there is no motion in the vibration and it just sort of pulses between here and "somewhere else" at the same time according to the uncertainty principle.

Tamashii Light

Everyone knows that if you put two frequencies together, the combination is more than the two frequencies. It also contains "beat" frequencies that are associated but different that original two. According to the Tamashii Model, collisions of these two "different frequency" particles do the exact same thing. Interactions can cause secondary "Beat" frequencies to be generated. This "beat" frequency is the photon emission [light]. So, in general, the close proximity of two different frequency particles is what causes light.

Just think of light as an illusion of vibration. Remember that you can't really see light. All that the brain gets fed is the vibrational content of these photons. The brain changes the frequency of the photon vibrations into colors and the abrupt change of color allows us to distinguish things.

Quarks

Changing the frequency can change the characteristics and union possibilities. What this means is that the huge particles we previously thought of as building blocks of matter, atoms, are made up of still smaller particles that can change because of some outside influences. This "change" can and often does include invisibility or the output of this light stuff. As we get inside the atom, if 3 or more of these tiny, tiny particles, called Bosons, join, they become one of 6 different types of Quarks [Charmed, Strange, Up, Down, Beauty, or Truth]. These are weird names for weird particles. Odd quantities of Boson particles are visible and appear to have mass, even numbers have no apparent mass because they are electrically neutral and don't tend to interact outside themselves.

Gravitons

One such particle is the graviton, which has no apparent mass, but has a gravity, which requires mass. This "theory" ties together photons, wave theory, particle theory, gravitation without mass, and these new discoveries of molecules changing characteristics. We at least have a chance at having a real definition of particles with the Tamashii Model rather than the usual "definitions without details" we tend to accept. When I say "usual", I mean when a text book simply states "therefore it can be concluded" and you read before and after the statement and there is no way that the result can be concluded except by defining away many exclusions and exceptions.

A photon is NOT sometimes a particle and sometimes a wave like you were told. According to the Tamashii Model, the quantity of particles collected by the photon and their frequency characteristics make them invisible or visible depending on what they come in contact with.

Let's say there are 2 fermions [quasi-mass] things minding their own business and they happen to be vibrating at different frequencies. What happens is that there is a beat frequency produced and the particles become visible in a string with that vibrational pattern, but there is something left over. If 2 frequencies beat, there is a negative and a positive beat frequency; one becomes visible as a particle while the other generates electro-magnetism or a PHOTON. The diagram following shows how this function might be interpreted in a "time-space" way.

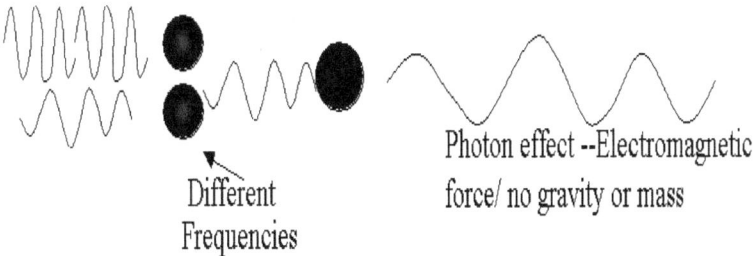

Different
Frequencies

Photon effect --Electromagnetic
force/ no gravity or mass

Tamashii Particles

Don't get me wrong here. The Tamashii model is not the total answer when trying to determine something about matter and photons. It does not define light by itself, but it does open up our minds to possibilities necessary to continue. All we have to do, according to this model, is look at the different particles and we will know everything, but they are really small. Some turn into light, some exert gravity, and some have no mass. Presently these particles are called bosons. Three bosons combine, they become quarks and depending on how they combine, the quarks act differently. Scientists have classified many of the quarks and given them very impressive names. Multiple quarks make other characteristic particles on-and-on until the atom is finally established. I came up with a short list of "defined" particles so that you would not get any more confused than you already are. They are separated as Fermionic [simplest sub-particles], Bosonic [Sub-Particles made from 3 or more Fermions], and Baryonic [Particles made for 3 or more Bosons].

Fermionic Sub-Particles

These are the "not quite particle" set. A fermion is a particle that has no apparent mass and exerts gravity and/or electro-magnetism. Previously entire particles like Electrons and Protons were considered fermions, however, most, today use the fermion name as a particle missing a component needed for it to really exist like a graviton. It has gravity, and must have mass, but no one can find its mass. All of these "quasi-particles vibrate at much lower frequencies than visible particles and all are made from Einstein's Aether [potential for matter].

- *Neutrino-This is a quasi- Lepton that is a component of an Up-quark, three types known [Electron-neutrino, Muon-neutrino, and Tau-neutrino.] They have almost no reaction with matter and can pass through the Earth--- They have no apparent mass.*

- *Electron hole-A lack of electron in a valence band*

- *Photon-This is sometimes considered a Boson that has no apparent mass so it really is a fermion but has electromagnetic properties. It can be modeled with 2 quarks or equivalent particles. It exhibits no Gravitational force, but instead it makes light.*

- *Graviton-Like the photon, this is another fermion possibility that has no apparent mass, yet it exhibits a strong gravitational force. This suggests that an even quantity of quarks is combined in its makeup.*

- *Chargon-A quasi-particle produced from an electron spin-charge separation*

- ***Configuron****-An excitation in an amorphous material associated with breaking of a chemical bond*

- ***Exciton****-A bound state of an electron and a hole*

- ***Fracton****-A collective quantized vibration on a substrate with a fractal structure.*

- ***Holon****-A quasi-particle resulting as a result of electron spin-charge separation*

- ***Libron****-A quasi-particle associated with the motion of molecules in a crystal.*

- ***Magnon****-A coherent excitation of electron spins in a material*

- ***Majorana fermion****-A quasi-particle equal to its own antiparticle in superconductors*

- ***Phason****-Vibrational modes in a quasi-crystal from atomic rearrangements*

- ***Phonon*** *-Vibrational modes in a crystal lattice associated with atomic shifts*

- ***Plasmon****-A coherent excitation of a plasma*

- ***Polaron****-A charged quasi-particle that is surrounded by ions in a material*

- ***Polariton****-A mixture of photon with other quasi-particles*

- ***Roton****-Elementary excitation in superfluid Helium-4*

- ***Soliton****-A self-reinforcing solitary excitation wave*

- ***Spinon****-A quasi-particle produced as a result of electron spin-charge separation*

Bosonic Particles

These are the base particles of matter. A little higher frequency that fermion, but still not making matter.

- ***Boson****-There are several different kinds of this particle. Some affect gravity, some effect magnetism, some effect nuclear forces. One theory is that these occur when matter and antimatter match and come in contact with one another. They sometimes exhibit no apparent mass but still produce a magnetic, gravity, or other force. The Photon and Graviton are sometimes considered bosons.*

- ***Gluon****-This is a boson and has no mass. It is a particle that can cause an interaction between Mesons. It exhibits a strong nuclear force. These are passed back and forth inside an atom's nucleus. There are 8 different types of these particles known.*

- ***Lepton****-Bosons that do not attract nuclear force. The electron is the simplest form. There are 12 known varieties of this particle with 6 of them having no mass. [Yes, Leptons without mass may be considered fermion.]*

- ***Tau****-This is a heavy Lepton that is a component of a Beauty-quark.*

- **Muon**-*This is a heavy Lepton that is a component part of a Strange-quark, this one is found in Cosmic rays so stay away from Muons.*

Baryonic Particle

These are the combined particle sets that build an atomic cloud. Still higher in frequency these are much bigger, but, again the frequency will still have to get faster to build the full atoms as was listed previously.

- **Baryon**-*This is a particle with 3 quarks, so it's usually heavy. The only stable Baryons are protons and neutrons. They have strong nuclear force. There are at least 10 unstable types besides the proton and neutron.*

- **Hadron**-*Quarks, mesons, and Baryons that attract nuclear force end up being called hadrons.*

- **Electron**-*A huge quantity of particles [3 full quarks] in a mass. This is the smallest stable particle.*

- **Proton**-*This particle is made up of two up-quarks, and one down-quark, because an up-quark is much more positively charged than the negative charge associated with the old down-quark, the proton looks positive. This is a Baryon particle.*

- **Neutron**-*This particle is made up of two down-quarks and one up-quark. Because the down quark is half the charge strength of the Up-quark, the particle is electrically neutral. This is a Baryon particle.*

Common Material Frequencies

Besides trying to determine what sub–particles were, the Tamashii model established what true particles and elements were. Next are a couple of tables that show the actual or theoretical frequency and wavelength standards of common elements known today along with other attributes of other characteristics of our universe. The material frequencies have been derived from the various groups investigating vibration reaction of structures/atoms. How would you like some particles vibrating at 60 exahertz? That vibration causes Gold, as you can see from the list following. Have the right frequency or resonate the environment around a substance and one can make the material you want. Notice that most frequencies do not form matter, at least structures with mass. Even the smallest physical component [BOSON] must vibrate fairly fast [300 MHz] so one would think that if you wanted to modify particles, you had better have a source that can vibrate very, very fast. One thing to note as you look at the tables; vibrating frequencies that create the element we call Meitnerium can even vibrate faster to produce the limits of the structural dimension or what we call pure magnetism. Some call it a black hole. It is known that the event horizon of a black hole can take dark matter and convert it into much higher frequency "visible matter".

Chart of Particle Vibrations

Name or characteristic	Maximum Wavelength [meters]	Highest Frequency [Hertz]
Aether [??]	$*1 \times 10^{+10}$	$<30 \times 10^{-3}$
Fermion [part mass]	$*1 \times 10^{+4}$	30×10^{3}
Boson [smallest mass]	$*1 \times 10^{-0}$	30×10^{7}
Baryon [electron]	$*1 \times 10^{-3}$	30×10^{10}
Hydrogen/1	1×10^{-9}	30×10^{16}
Berylium/9	1×10^{-10}	30×10^{17}
Silicon/28	3.5×10^{-11}	8.5×10^{18}
Zirconium/91	1×10^{-11}	30×10^{18}
Gold/197	5×10^{-12}	60×10^{18}
Meitnerium/270	3.7×10^{-12}	27×10^{19}
Straight Gravity	smaller	higher

Why Discuss Matter Vibrating?

I suppose you are wondering why I bringing all this stuff up, but I need you to understand that everything can make photons. When we get to how DNA produces photons and how <u>cancer is fed by photons</u>, and <u>miraculous healing occurs from photons and optical communication of sorts</u> seems to happen between cells in a body and in plants. Before I get into this very important section about light, let's first get a picture of the stresses in our universe as frequencies of matter, living organisms and photon streams being viewed at high rates of speed.

If the amount of mass changes in this universe where does the mass go????? If the composite frequencies increase to make more complex particles, where does the increase actually come from????

That's where we see how important a linked "regulatory" universe is in assuring our universe stays stable. We see photons being emitted all over the place and the vibration frequencies of matter are changing as materials change. Unfortunately, there is a law that says one cannot create matter, time, vibrational average or anything else. There must be a way to make mass appear allow these other things to be modified. This is especially problematic when matter tries to increase in frequency and the LAW OF ENTROPY tells us everything WILL attain its lowest order---or lowest frequency. Without an adjacent universe, soon we would lose all matter and become something less than matter. I know that makes you sad, so they found out about Super Symmetry.

Super Symmetry

Super Symmetry is the Theory that if a frequency is reduced here, it is expanded in an adjacent universe in an equal and opposite way. This concept may help us with the anomalies of Photons and light.

Everything is symmetric by constant mass. For light, when the "energy" disappears here, it is immediately sensed in a different universe.

That is, a large particle has a sister particle that is tiny while a tiny particle has a sister very large particle. The same thing holds true for Light. Examples are given below:

- Photons have duals called Photinos
- Quarks are partnered with Squarks
- Matter has antimatter
- Electrons have positrons

No, I did not make up the stupid names! Have you had your squarks today?

So long as the particle twins have the same mass as other particle twins, all is fine in both universes. Everything is symmetrical; well, almost everything is symmetrical. <u>If we have a life in this universe, it does not mean we have a death in an adjacent universe. There is simply a different type of life, but that is for another day</u>. As the law of Entropy works in a linked universe, we get an advantage of higher frequencies which means MORE matter. Sometimes we have to give up matter as well so everything stays regulated. We don't have to worry about the Law of Entropy and we can manipulate photons as needed.

Changing Matter

Just like moving away from something changes its characteristics because the vibration "effect" changes [Dr. Hubble's Red Shift], by simply changing the vibrational base of a substance, the material changes. If you want to change lead into gold simply change its characteristic vibration and it becomes gold. One way is to simply have the lead go away from you fast enough. With electro-magnetic force, changing the vibration level changes the Photonic material from radio waves to cosmic Rays. Hopefully you are seeing a duality between light and perceived particles.

Scientists today are changing lead into gold by forcing protons out of what we call nuclei in atomic clouds. Making the nucleus smaller makes the frequency of vibration slower and you have just changed matter. Another way is being accomplished by a man named John Hutchison. He has done the seemingly impossible by making metal invisible, changing hard objects into pudding, and having solid objects go through solid object without being seen. Generally, this Hutchison effect can be called vibrational Invisibility.

Vibrational Invisibility

We know that if you have a vibrating object vibrating completely out of phase with another visible object, they vanish. This is why gravitons have gravity, but no mass. While there must be a mass there, it is invisible in our universe. This is also the reason that what we call light sometimes has a mass and other times has no mass. Before we look at some of Mr. Hutchison's discoveries, let's look at his predecessor named John Keely.

John Keely

In the 1880s John Keely possibly came close to the truth about light. During his investigations, he invented numerous devices that seemed to channel vibrational energy, so this seems like a reasonable starting point for inventing light. He claimed to have discovered a new motive power which was originally described as "vaporic" or "etheric" force, and later as an unnamed force based on "vibratory sympathy", by which he produced "interatomic ether" from water and air. John is shown below.

Vibrational Sympathy

He refused to reveal the secrets of his inventions and methods. In 1884, his "Vaporic gun" was demonstrated. His description of what it did is very interesting to use. He stated, "*I take water and air, two mediums of different specific gravity, and produce from them by generation an effect under vibrations that liberate from the air and water an inter atomic ether. The energy of this ether is boundless and can hardly be comprehended. The specific gravity of the ether is about four times lighter than that of hydrogen gas, the lightest gas so far discovered."* [New York Times, 22 September 1884]

Certainly, he had some things wrong, but he was on the right track by investigating the vibrational component of structure. OK! He mostly put a bunch of long words together, but it's the vibration and ether things that are of interest to us here.

He also indicated, *"It is an elaboration of interatomic ether by vibration. The atomic ether vibrates all around the molecules of matter. There is a magnetic force attached to it at the same time, and it assimilates with the molecular atomic aggregations - that is, assimilates with a certain attractive force that it is hard to tell what it is. I call it a **vibratory negative**. It doesn't act like a magnet drawing metals toward it. There is a certain magnetic effect about it that causes it to adhere by vibratory rotation to different forms of matter - that is the molecular, atomic, etheric, and ether-etheric. The impulse is given by metallic impulses, the rotary power that is formed by etheric vibration - that is the force that holds it in position."* [New York Times, 7 June 1885]

Vibratory Negative is a very interesting concept. Just think about it. We now know that if two vibrational components are put together out of phase, they disappear. One is the vibratory negative of the other.

This is how light can be self-generating and how it can simply disappear.

Vibrating Subatomic Substances

Keely's Law of Vibrating Subatomic Substances also adds light to our investigation. *"Atoms are capable of vibrating within themselves at a pitch inversely as the local coefficient of Gravity, and as the atomic volume, directly as the atomic weight, producing the creative force, whose transmissive force is propagated through subatomic solids, liquids, and gases, producing induction and the static effect of magnetism upon other atoms of attraction or repulsion, according to the law of harmonic attraction"*. [He had correctly understood the perpendicularity of gravity and mass vibrations and their affect.]

Oscillating Subatomic Particles

Keely's Law of Oscillating Subatomic Particles gives us still more information. *"Subatomic particles, oscillating at a uniform pitch produce the "creative force", whose transmissive form, "Gravism", is propagated through more rarefied media, producing the static effect upon all other sub atomic particles, denominated as gravity."*

Vibration Definition

Finally Keely's Definition of Vibration helps us some more. *"Life in its manifestations is vibration. Electricity is vibration. Vibration that is creative is one thing. Vibration that is destructive is another. Both may be from the same source."*

He had somehow discovered that the Spirit force in the Ethereal Dynamo <u>was affected by the same type of vibration as that found in electricity</u> and he had not even read my book on the subject.

Keely's Inventions

John Keely's inventions were endless and finally he was labeled a charlatan but his work should not be discounted as he ventured into a region that could have enlightened many if they had only kept their minds open. It would be a long time before John Hutchison would come along.

The Hutchison Effect

If you haven't heard of this man, shame on America for not teaching about his work going on in Canada. Using combinations of electromagnetic waves, John bombards items to see what happens to them. Objects go nuts. Not to be humble, the effect was named the Hutchison effect. It seems that the effect is not always the same.

Nucleonic Energy

The Hutchison Effect has sometimes been called Nucleonic Energy as identified and classified by another Canadian named Dr. Mel Winfield. For those interested in the details of his unified particle theory, it's pretty amazing stuff. In Winfield's book, *"The Science of Acuality"* for instance, he defines gravity as *the differential velocity between central core elements of a substance with higher rate objects on its outer surface.*

Yes, John Keely said something similar in the 1880s and some form of this effect was actually discovered and called the Magnus Effect, but Dr. Winfield refined and redefined the elemental parts to allow us to understand the workings of **gravity, levitation and vision a little better**. So long as the electrons flew around the nuclear mass at a rate faster than the nucleus, gravity happened and the electrons and, for that matter, the nucleus, stayed together. Each electron banding spins faster than the inner grouping forcing those "electrons" in an orbital around the nucleus. Simply reverse the nucleonic spin and things go flying. Things start levitating. By phasing the oscillations just right, the nucleonic energy is not sensed by the surroundings. It becomes gravitationally invisible.

Levitating objects was possible by phasing the oscillations of a material.

Mass Invisibility

If two vibrations come in contact with one another, One being associated with a heavy object under test and a second one, possibly an electromagnetic vibration that is identical to that characterized by the object, some pretty strange things can happen. If they are phased such that they are exactly opposite in their vibrational cycle, guess what happens? The next example shows what one would expect.

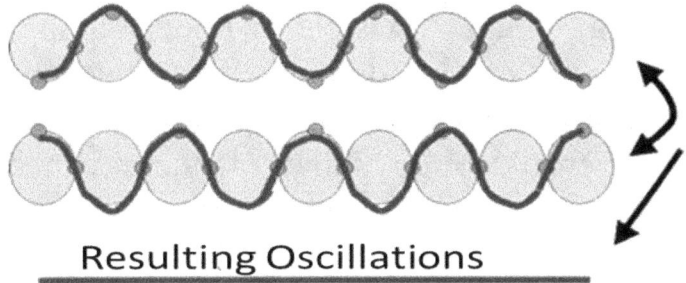

Resulting Oscillations

To the viewer, seeing both vibrational constructs, there would be NO vibration. With no vibration, the object would cease to exist until on electromagnetic field was eliminated. With different phasing, the object can be made to be weightless, or even change its characterization and John Hutchison, accidentally found out how to do it. When I say that, the task is not so simply accomplished as heavy objects have a vibration frequency that is very, very, very high. Gold vibrates at about 60×10^{18} hertz [60 Exahertz] and no electronics on earth can transmit something like that.

Levitation Effect

While seemingly impossible, for John, "sometimes" items levitated when blasted with a **wide assortment of electromagnetic frequencies**. I'm not talking about a paper clip spinning around on a table by the force of a magnetic field here. I'm talking 50-pound bowling balls and heavy objects. They became "Invisible" to the gravity of the earth because the particles that made up the objects were vibrated just at the right frequency.

While John Keely and Dr. Winfield were both radical thinkers, we cannot ONLY talk about modern levitation and making things invisible without John getting involved. He not only made things invisible to magnetic fields, he made thing completely invisible. You have to be a little crazy to do some of these experiments. Remember, John Hutchison is a man experimenting on these dangerous aspects of physics in his own home. John Hutchison has been conducting some amazing, well witnessed, experiments and, I'm sure he will, soon, find out what produces the "Hutchison Effects" that are beyond the initial gravitational invisibility described in "The Science of Actuality". Right now let me just show just a tiny grouping of his levitation successes in the presence of a powerful and strange field of electromagnetic waves. Pliers are yanked into the air, a bowling ball hovers.

Various plates fly like planes [shown next]. It almost like a ballet, but how in the world is it possible?

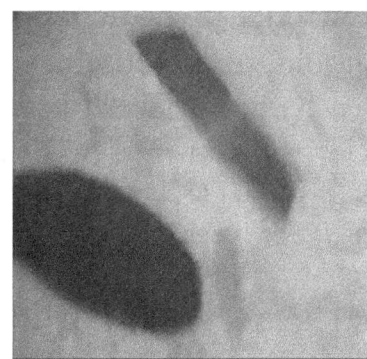 

Metal Turned To Jelly

Sometimes metals change consistency and become jelly like before returning to metallic state in a stringy mess as shown below.

 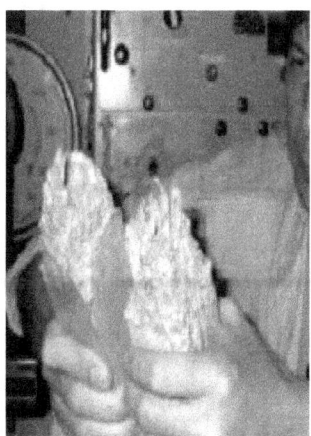

Objects Go Through Metal

This phenomenon is the one that will help us understand invisibility. Sometimes, objects started floating and went right through metal objects. If the process is halted before the material can go all the way through, the object traveling through the metal become part of the metal as shown next. The first picture is of a butter knife that began traveling through a metal plate. It ended up integrated with the metal. It should be noted that the crystalline structure of the metal was not affected by the intrusion. The second picture is of a piece of wood that was caught as it traveled through a plate of aluminum.

Invisibility

Possibly the most important thing for our study of light is that objects would become invisible. Sometimes objects began to fade away and finally they became totally invisible. The following images are a couple of images during the transition.

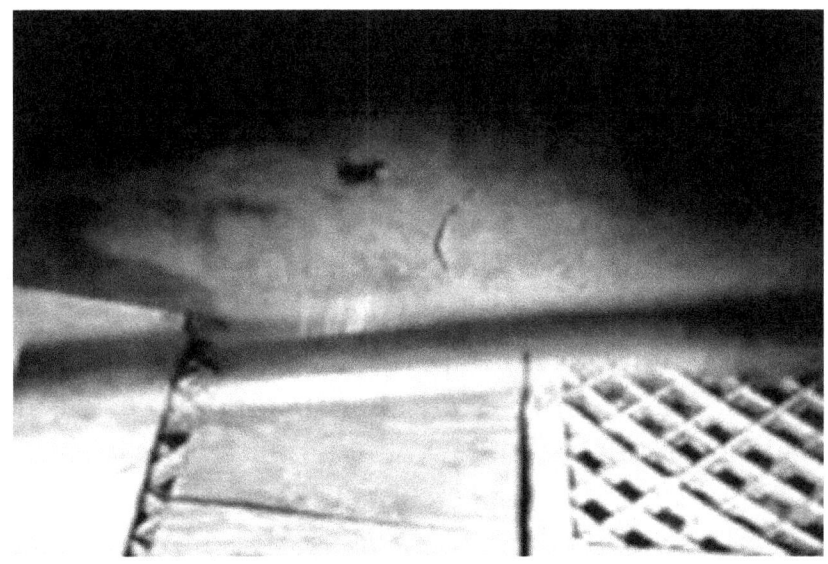

Note how you can see through the tube above and how the wood can be viewed through the metal bar [next], only when the electro-magnetic fields were bombarded towards the objects.

Explanations

Let me try to explain what is happening. Let me first say most people simply scratch their heads. I hope you aren't doing that, because it is really neat and will help us understand what light is. When I said there was nothing on earth that could generate the type of frequencies that would allow for these things to happen, John did it just the same by combining a number of high frequency generators and synchronizing them with very slight offsets in phase. This is an almost impossible task and needs substantial patience. John simple kept moving his dials to change frequencies and viewing the results. When he hit upon the almost phase locked frequencies, something used in noise cancellation took over.

Noise Cancellation

Remember noise cancellation? It works by adding noise. You would think adding noise would make it noisier, but if the noise is out of phase, the signal is neutralized at the ear. John had the microwave generator all outputting almost the same thing and the outputs combined to make harmonics. Harmonic of the harmonics make higher frequency harmonics which can combine to make even high frequency harmonics.

If you get it just right and phased just right such that your emission is out of phase with the part that is irradiated, its "ESSENCE" of vibrational character can be neutralized. With matter, this doesn't just make the matter invisible; it makes it cease to exist for a time.

Light Example

Just like noise cancellation, light can be canceled by irradiating the same area with out of phase transmission of the same color. All of a sudden—there is NO COLOR. Does that mean that even though we know there is color generated that it is present?---- I told you defining light wasn't going to be easy. Well; even defining photons is going to be difficult.

Photons

We started looking a light and the scientific powers announced that light was sometimes a particle and sometimes it was something called a wave. While that is a whole lot of malarkey, let's look at this wave thing. What these guys said was that light sometimes doesn't exist as a thing. Instead, a whole bunch of nothingness was vibrated faster and faster until people saw stuff. Still faster the nothing vibrated and people could see through people. If these photon things still vibrated faster they, collectively, are called "cosmic rays" which started killing everything in "their" path. OOPS! I can't identify cosmic rays as a thing. "They" are defined as nothingness vibrating which can't be in a 3-dimensional world.

Take a minute for all this stuff to sink in. When you are ready, continue!!

Next is a chart of the thing we **<u>incorrectly</u>** call light.

Description	Cyclic period	Frequency	Feature
Helpful Infrared light	1×10^{-6}	30×10^{13}	Invisible thing
Visible light	4×10^{-7}	75×10^{13}	Visible thing
Dangerous X-rays	1×10^{-8}	30×10^{15}	Invisible thing that penetrates bone
Deadly Gamma Rays	1×10^{-9}	30×10^{16}	Invisible thing that destroys

If this was light, we could see it---right? Let me tell you a secret even the part that is called visible light is not always visible.

Electro-Magnetics

Someone said, "What if we slow this light stuff down; can we make radios work?" Wow! There it was, slow vibrating light was making radios work and the study of electro-magnetics brought us AM, FM, Televisions and the internet all without any of these slow photons getting MASS so that they could be defined in a 3-dimensional world. Below is an expanded "light" chart.

Name	Maximum Wavelength [m]	Highest Freq. [Hz]
Pure DC Voltage/ electric potential	5 x 1012	~0
Human hearing*	1 x104	20x103
VHF [radio]	1 x100	30 x 107
UHF [radio]	1 x10-1	30 x 108
SHF [radio]	1 x10-2	30 x 109
EHF [radio]	1 x10-3	30 x 1010
Microwaves	2.5 x10-4	12 x 1012
Infrared [light]	1 x10-6	30 x 1013
Visible light	4 x10-7	75 x 1013
X-rays	1 x10-8	30 x 1015
Gamma Rays	1 x10-9	30 x 1016
Pure Magnetism		Really fast

*Note on the chart- The human hearing frequencies are not normally thought of as electromagnetic, but what I'm talking about here is the feelings one gets from certain sounds or how those frequencies affect thought and brain function. Lower frequency sound puts a darkness in our images and faster vibrations bring in more light so there is some type of connection.

What is the difference between mass and energy? I know you are going to say frequency, but, matter vibrates at the speed of light just like electromagnetics [EM field]. There are 2 differences, EM fields are going backwards in time with respect to matter and they travel in a straight line, while matter travels in a circular world going nowhere. When I said matter was going forward in time, I did not mean anti-matter. That goes backwards in time just like EM fields. Anti-matter simply doesn't belong here. It belongs in our linked universe where the EM fields originate.

The next 2 things I'm going to discuss are going to get sticky, but it is needed to convert EM waves into Sight and Light. The first is how time travels and the second one is the basis for Anthropic the collective of consciousnesses make our reality meaningful.

Time Vectoring

I know that some of this is getting hard to understand so I'll try to draw pictures. The matter and electro-magnetic force components of the universe are completely <u>inertial based in that all emanations can be characterized spherically</u>. Unfortunately or fortunately, time seems to act adjacent to the other key elements of matter and force. While forces all must be identified with time and waves must be characterized with time, we now know that this time component is not constant in an absolute. <u>Time is relative</u>. While one could say that means that the structure of the universe would expand or contract depending on the time constant used for the model. It is this dependency on reality that makes time so hard to introduce. Placement of all of these wave nodes starts shifting and <u>confusion ensues</u>. Time, however, seems to react to velocity. While that seems like a self-fulfilling element as velocity only has meaning with time, the thing to recognize in Einstein's model is that time is "Vectorized" as shown next in a picture I drew.

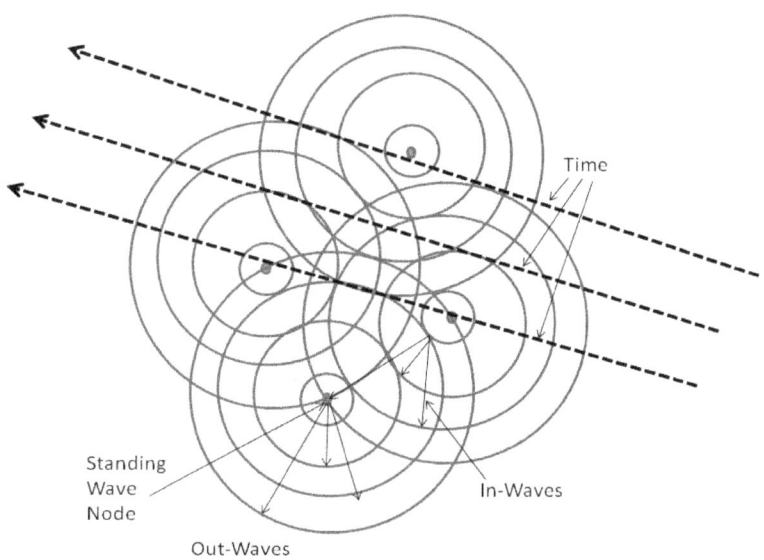

Time

Standing
Wave
Node

In-Waves

Out-Waves

Time stays constant to an observer. Wherever he goes and however fast he goes, **<u>"Time" stays with him</u>**. That is he experiences everything exactly the same if he goes close to the speed of light or if he could possibly not move through space at all. Experimentally, we show that along the axis of a velocity, mass increases. If we could see a table, for instance, the table would get longer in the direction of its velocity vector. NO! NO! NO! In the relative world the table stays the same spatially and the forces [all of the in-waves] react to cause the same forces to the traveler and the objects moving with him. Time can be considered as a rectangular element of the polar universe. Straight-line augmentations of the fabric directed to a vibration. Because none of us are pulling on this fabric the same, the universe must be characterized as multiple universes all being vector driven by time differential associated with time.

Relativity Helped with Anthropics

We need to add something else to the model or all of the relativity, and Anthropics go out the window. Hopefully, my drawing below, sort of, shows how all this fits. While matter and forces are affected by time, the collective consciousness of cognizant viewers cannot sense it, but it must exist to address relativity, so it is perpendicular to both Mass and Force and perpendicular to the effects of time. I drew it 2 ways to, hopefully, describe it a little better. Some may call this the soul or soul collective.

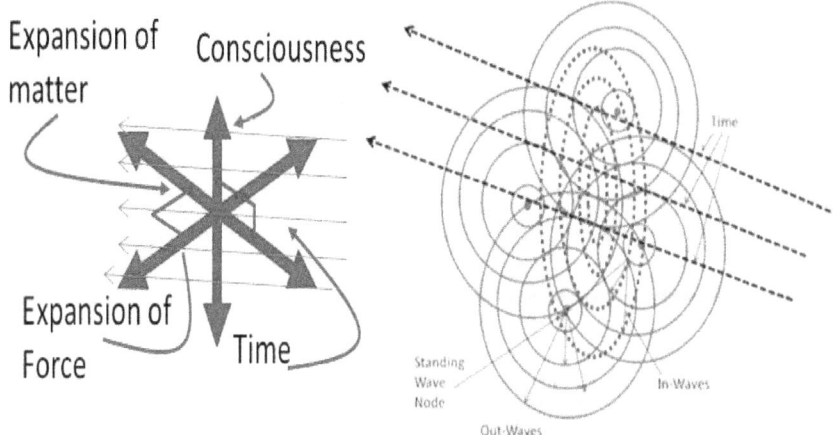

Soul Description

The meaning of life is not 42 as indicated in a popular science fiction film. Instead, it is a complex variant of our universal existence. It is said that if a tree falls in the woods and no one was around, the incidence simply does not happen. According to the science of Anthropics, without a cognizant observe, the universe modifications cannot take place. One way to tackle this subject is connecting life to time. That is, time without life or vice versa means little. More precisely, connecting time and the conscious mind [soul] must be done to characterize the universe that changes by perspective in Einstein's relativity theory. There is no light if there are no people. They are both undeniably attached and work as a single unit. That is not to say the soul is time because cognition seems to be "timeless" or accomplished adjacent to time. This is saying that the conscious mind is projected into the universe perpendicular to time so it does not place restrictions to time.

Because the vector is perpendicular to the various velocity vectors and abstract to the in and out waves, this soul dimension is not characterized in the normal sense. You could say it "usually" doesn't exist. String theorists would say the dimension is compactified. I would simply notice that the soul has complete freedom with respect to time. Because the soul has this freedom, people have been able to sense the future and see the past. People have been able to lift automobiles off people who were trapped without crushing their own bones. People have been able to walk on water and turn sticks into snakes and all the rest identified in our most holy historical references.

Now I will ask. If you see 700nm photons and someone else sees 700nm photons do you see the same thing? Common sense would suggest that how the brain processes the various wavelengths of electromagnetic pulses would have to be different, but if consciousnesses are linked, when I see red, others see the same thing. Don't let people come up to you and tell you that you simply define red one way and they define red another so that we can talk about wavelengths in a similar way------- - The answer is that we are all attuned to this changing reality and we are responsible, in a small way for changing our perception of reality and the color red. It doesn't matter how fast we go or how strange someone is, Red is not just defined, it is established by our collective consciousnesses.

How Fast Are You Going?

I know I briefly described this already, but it may help us understand this Anthropic view of reality. If you live in Florida, you are spinning in a circle about 1000 miles an hour already as the Earth rotates. Because the direction is across your body, you are fatter on the equator than at the North Pole where you would go slower. If that isn't enough, we are spinning around the sun at a rate of 66,000 miles per hour. Besides that, our solar system is milling around in the Milky Way at a rate of 43,000 miles per hour. Now for the bigger numbers the Milky Way is rotating such that it takes 225 million earth years to make a galactic rotation. This means that the galaxy is turning at a rate of 483,000 miles per hour. We certainly aren't done yet as the universe appears to be expanding. While expansion is an arbitrary term, red-shift analysis tells us the Milky Way Galaxy is moving at a speed of 1.3 million miles per hour. We are moving roughly in the direction on the sky that is defined by the constellations of Leo and Virgo.

That all being said you are moving at least 1.3 million miles an hour so how fast is light to you? 670 million

MPH just like it would be if we weren't moving at all.. Someone seeing the light that you are shining that is "stopped" in space sees a light shining at 670 million MPH as well, but you know in your heart it must be going another 2 million MPH faster, but it is not! This is where light and photons separate. Electro-magnetics is constricted by the in-waves and out-waves of the universal sphere. But light stays constant to a conscious entity because consciousness is one of the dimensions of what we can call our "personal universe".

There is a change in the light, however. To the stationary observer, the number of cycles a photon vibrates while it is moving is less than experienced by the person on the earth moving so quickly. To the stationary guy, there is a shift towards the slow light oscillation levels or towards the red side of the visible spectrum. It's like the photon stream is stretched to keep both viewpoints seeing light properly in "their" reference point. I know this seems bizarre, but this has been proven so please understand that your consciousness affects light.

Let's Go Faster

Just for kicks, let's see what the light will look like if the difference in speed is really, really high. Let's say the guy shining the light in the last example is going close to the speed of light himself. The apparent wavelength would get longer and longer and appear to be radio waves so we couldn't really see a light shining at all.

The thing to understand from this section is that light is a personal thing in your personal universe, while photons are shared-- light is not.

If you don't learn anything but the above, you will be miles ahead of most people today. Don't worry about right now. I will get into what that means in more details as we get into Anthropics more.

Photo-emissions of Cognizant Observers

If I told you people send out photons from their cells would that seem odd? I'm not talking about Kurlian photography where odd images were photographed showing what had been defined as an aura as shown below from hands, leaves freshly picked and metal objects. Let me tell you the metal is not alive so there is something else going on. The images below are some that are typically shown.

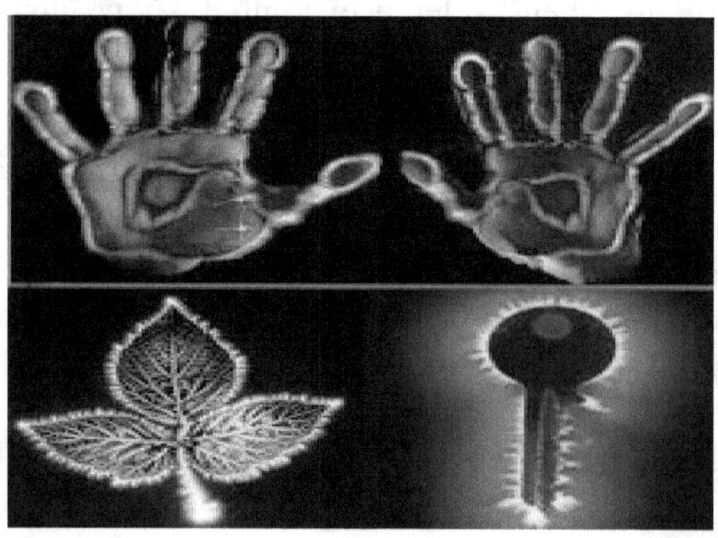

What I'm talking about is more unusual, and important in our finding out about photons and light. What we now know is that all living entities emit photonic instructions. This is so important that it could soon be the cure for diseases like cancer and allow normal regeneration of cells. It may allow us to grow things more efficiently and it is all done with photons. This is turning into a brand-new science of light that is forcing us to understand photons and light better. The next few sections deal not with the esoteric elements of what light is but how is it associated with DNA and live structures. We already know that photons and matter have a strange commonality and structure. Depending of wavelength, electro-magnetic waves can modify placement of electrons and even entire atomic cloud placement. Photons can even become mater. What scientist are finding out now is that living cells talk with light.

Practical Application of Biophotonics

Bio Photons

It has been known for some time now that our body, all other animals and all plants transmit and receive photons. Depending on how they are sent, things change in nearby cells or even nearby people or plants. We had better find out about this or we will never know what photons are. While the majority of the emissions are in the Ultraviolet range, some are visible. The levels are low as the photon communications are for short distances.

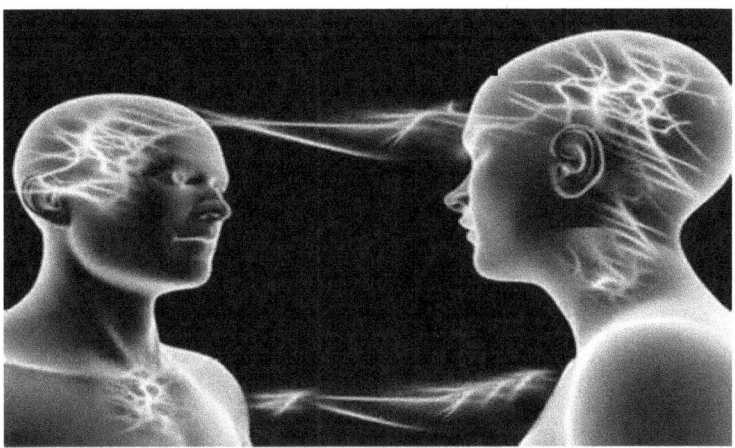

The image above might show how feelings and image transfers might look if we could see them. I think the best example is when someone cuts their hand open. The body needs to repair itself, so the affected cells send out a distress photonic message to the nearby "good" cells" that they need to replicate. The cells get that message, replicate, and, soon the cut is completely gone. The messages are by pulse coding and differences in output wavelength. The most complicated and manipulative bio-photonic generator and receiver is in the molecules known as DNA.

DNA Emits, Detects, and Decodes Photons

One of the leaders in this study of light is a Russian scientist named Pjotr Garjajev. He recently was able to intercept UV Photonic communication from a DNA molecule from one organism, a frog embryo, and retransmit it to another organism's DNA, a salamander embryo, causing the latter embryo to develop into a frog! Evidently what happens is the DNA sends out a message that can be seen by cells passing nearby, if they receive a particular optical message, they might turn into skin, or whatever. The main thing is that the old idea that everything was accomplished by chemical modifications making electrical differences that were interpreted by cells has now been thrown away. Light builds people, animals, and plants.

Plant Cells Emit, Detect and Decode Photons

A Russian biologist was the first to find this out. His name was Alexander Gurwitsch who experimented on onion cells and found that stimulating one onion that was near another would cause the second one to flourish if there was quartz between them but not if silicon was between them because the biophotons that were being transmitted were Ultra-violet. With no barrier or quartz, one onion being fed would cause another to react. That whole concept of talking to your plants is gone------ now you need to send the right photons and they need to be the UV light. After the Onion Tests, I figured <u>hospitals were not the place to be if one was sick unless people were placed in separate rooms or something that would halt UV light separated sick people.</u> Plants would call out to those nearby not with simply pheromones but with light. The image below shows a cry for help when an insecticide is sprayed. The photonic emission shouts out the fear and hurt and soon settles down.

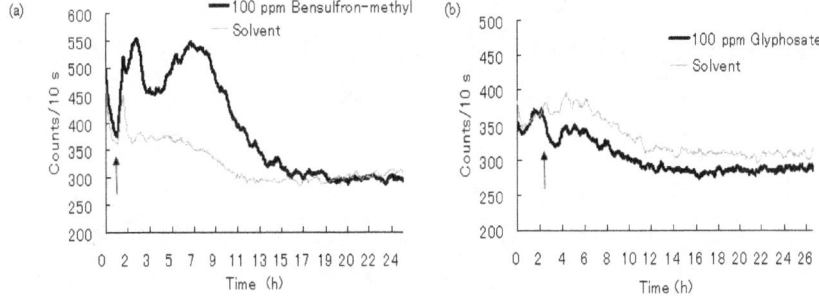

Photo-repair Sensing and Cancer

Here is something even more strange. It seems that scientists found that blasting a cell with 380nm UV light so that 99 per cent of the cell, including its DNA, was destroyed would not be the end of the cell. All you need to do is reblast the cell with a very low dose of 380nm and the cell will regenerate in a single day! I really don't know the exact nature of all of this nor do I believe anyone does. One thing I do know is this sounds like the rapid growth of cancer so let's see what happens there.

Hydrocarbons Detect and Emit Photons

Some believe this Photo-repair messaging may be the cure for cancer. Hopefully, by now you are starting to see how very odd photons really are and how linked they are to what we call reality. A theoretical biophysicist at the University of Marburg in Germany named Fritz-Albert Popp started his work in 1970, examining differences in a carcinogenic hydrocarbon named benzoapyrene, and almost identical but safe one named benzoepyrene. Like most experiments in the "talking to DNA" arena, UV light was the thing to activate photonic emission of these cells, He had illuminated both molecules with ultraviolet (UV) light in an attempt to find exactly <u>what made these two almost identical molecules so different</u>. The first one emitted a different spectro-metric wavelength pattern while the safe hydrocarbon reemitted the same pattern as was input. He found out that other cancer-causing hydrocarbons did the same type of photon change and they would only react to 380 nm. Once activated, massive output of photons erupted from cancerous cells to invigorate a change to cells nearby as shown below. This looks bad.

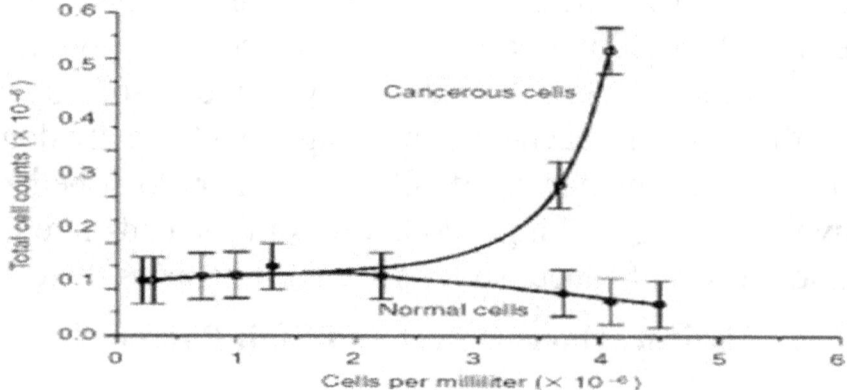

It was also noted that Melanin is capable of transforming ultraviolet light energy into heat such that more than 99.9% of the absorbed UV radiation is transformed from potentially genotoxic (DNA-damaging) ultraviolet light into harmless heat, but destruction of that safeguard would quickly allow 380nm signals to begin changing cells. I would have to say get as much of this melanin as possible on your skin.

Photons Control Everything in Living Organisms

It seems Photons switch on and control ALL the body's processes. Given different frequencies, identical cells perform different functions. The question that has forever puzzled cellular biologists for decades has been, *" What is it that enabled the tens of thousands of different kinds of molecules in the organism to recognize their specific targets?"* We now are beginning to understand how it is happening. It's not helping us define what photons are and what light is, but it is given us the details we need to begin to construct a definition.

What do Bio-Photons Do?

Bio-Photons produce or establish life. They explain how enzymes can recognize their respective substrates, how antibodies in the immune system can grab onto specific foreign invaders and disarm them. By extension, that's how proteins can 'dock' with different partner proteins, or latch onto specific nucleic acids to control gene expression, or assemble into ribosomes for translating proteins, or other multi-molecular complexes that modify the genetic messages in various ways.

It seems that <u>somehow</u> each molecule <u>sends out a unique electromagnetic emission</u> that can "sense" the field of the complimentary molecule. By this, molecules recognize their particular targets and vice versa by electromagnetic resonance. In other words, the molecules send out specific frequencies of electromagnetic waves which not only enable them to 'see' each other, but also to influence each other at a distance and become drawn to each other. With about 100,000 chemical reactions happening in every cell each second and each one initiated by some special coding of bio-photonic emissions, you can see that photonic energy is switching on and off continuously. One researcher put it this way---

"We are swimming in an ocean of light."

It has been suggested that the way DNA is coiled is to allow it to change its "resonance" to send and receive various frequencies needed to support life.

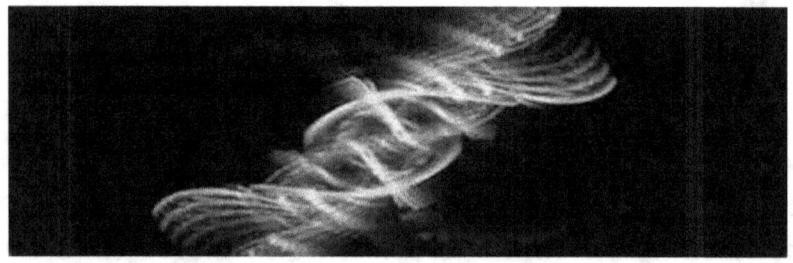

As shown previously, like a tiny spring, higher frequencies would have it tighten while low frequency emission needs would have it loosen its coils to allow better transmission. It is also known that as the DNA uncoils, the amount of light emitted is increased. We now can sense what happens when you get a cut or scratch on your skin. The images following show the extra photonic activity in areas of distress on our skin and anywhere else on or in our bodies. This is not just light. It is a message carried in the light.

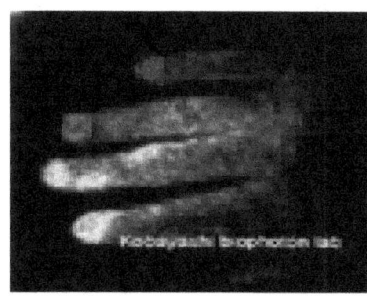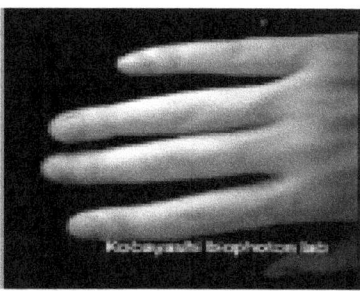

Now For the Halo

If you remember from Religious histories, there was mention of Bioluminescence or HALOs. Those holding substantial amounts of "Faith" would begin to glow with something called a Halo. When Moses came down off the mountaintop after talking with God, he had one of these halos as bio-photons were emitted in huge amounts.

Exodus 34:29-35*-Now it was so, when Moses came down from Mount Sinai--that Moses did not know that the **skin of his face shone** while he talked with Him. So when Aaron and all the children of Israel saw Moses, behold, the **skin of his face shone**, and they were afraid to come near him. And when Moses had finished speaking with them, he put a veil on his face. But whenever Moses went in before the LORD to speak with*

*Him, he would take the veil off until he came out; and he would come out and speak to the children of Israel whatever he had been commanded. And whenever the children of Israel saw the face of Moses, that the **skin of Moses' face shone**, then Moses would put the veil on his face again, until he went in to speak with Him.*

Today, scientists are finding out this is a truth that needs to be considered as very low levels of invisible and visible light are both emitted from the body and during times of stress, this output is increased. All organic life absorbs, emits, and processes light. Bio-photon emission or spontaneous ultra-weak light emission has been observed from almost all living organisms, with intensities ranging from 10^{-19} to 10^{-16}W/cm^2. A number of studies also found that emissions change by various cycles of the body. In Moses case, His head turned bluish. Possibly his whole body was the same, but most was already covered. The image following shows how his halo might have looked.

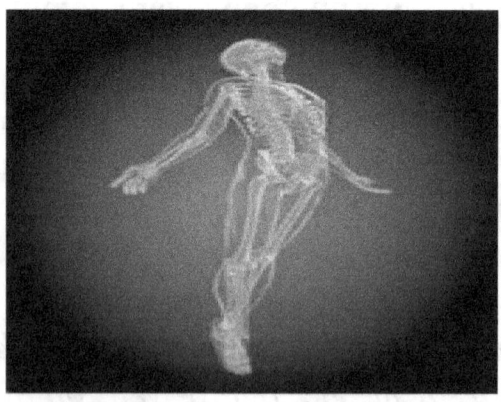

Bio-Photonics Has Taught Us a Few Things

While photons are a major dimensional element in the universe, in living entities, it can have additional impact beyond the all-out establishment of all forces we need to build a universe in that photons can help mold life itself.

- Light is associated with all life.
- Without light there is no life.
- We can make a general statement that Cancer is too few coherent biophotons and Multiple Sclerosis is too many coherent biophotons.
- Stress Increases light output [perhaps as a plea!].
- If plants were not stressed before being harvested, they will not output as much light.
- If one does not have major issues in his body, cancer etc., light emissions in the tested areas will be low.
- Fluctuation in photon counts over the body is lower in the morning than in the afternoon.
- The upper extremities and the head region emit most the most light and it increases over the day.
- Highest energy levels emitted by humans are 470-570 nm. This means that if one had enough people in a dark enough area and they were stressed, there would be a tiny blue haze that could be recognized. That being said, there are many wavelengths emitted by the cells as shown next.

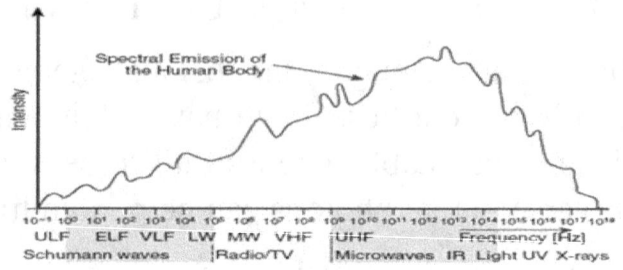

Common Brain

It has been suggested that biophotons connect us all to a sort of planetary brain by way of interactive travel, retransmission, and confrontation. It could very well be that these emanations are the catalyst for establishing the communal consciousness that drives our anthropic defined reality. As we feel bad, that bad feeling modifies light emissions which make very slight changes to reality. With Photons crashing into matter and changing it and photon emissions in living cells controlling the surrounding environment and the concept that Electromagnetic waves and matter waves are very similar and possibly even identical, I think you vision of the world has to have been expanded, but I need you to take an important leap of faith here and we get into a very important part of photons and light in our universe. We need to look at the Anthropic Principle and get a better understanding of how reality is being held together.

Given all the red shifting and frequency changing, what does the color red look like?

To someone not attuned to our collective consciousness, he would have a difficult time understanding the world around, but all of us see the same red and interpret it exactly the same. For that matter, our size is compared the same to all in our collective reality. Additionally, the various forces holding atoms together and allowing electronics to operate properly and all the rest are regulated the same because of our collective consciousness.

Dead People

While I'm on this odd subject let me just say, life cannot be created or destroyed ---that doesn't mean there are no dead people, but the soul must stay viable unless it can be replenished in some way.

Right now, the easiest possibility to understand is that souls don't die. That would mean that most of our reality is based on what dead people control.

I know some believe that the Bible is in error when it indicates we all sleep until the end of days and that the instant we die, we are whisked away to heaven, but that is not in the Bible. I know some are going to pull up the thief on the cross statement, but that is not the reason for this book so we are going on with the Anthropic Principal to get a better idea how all this works.

Participatory Anthropic Science

Anthropic Principle

I can imagine the dead people remark has gotten some of you upset, so I take it back. [That doesn't mean it's not something to consider, but I worry about you getting bogged down and not finding out about light.]

The idea that consciousness affects reality is not new, but now it has a sophisticated name "The Anthropic Principle" or the Anthropic Universe Theory". No one likes this idea because it makes the concept of time more difficult, but the whole Quantum Mechanics era has really destroyed our concept of stable time and stable reality, so, we are stuck with it no matter what. The whole "Power of Positive Thinking", "Think and Grow Rich" and all other concepts of the 70s which tried to convince you that how you consciously view reality will affect reality are not only true, they affect your death as well. In the Anthropic World if you have faith of a grain of muster-seed, you can move a mountain, as Jesus said thousands of years ago and you can walk on water as demonstrated by Elijah, Elisha, Peter, and Jesus did so many years ago. With the Anthropic Principle, science and religion can act as a single tool for us to understand God, the universe, and ourselves.

Before we can understand Anthropics, we have to look at quantum mechanics, I'm afraid. Speaking of quantum Mechanics, 2 different experiments have recently been able to transport information associated with a particle and regenerate the particle at a long distance without time delay. For Star Trekies, I said that they are now teleporting things using a quantum engine that eliminates space-time. The Anthropic Principle tells us that WE (all conscious life) CAN affect the physical characteristic of ALL things.

More than just transmitting light, our soul is part of reality.

The image below might show this crazy extension of our "self" into a cosmic collective.

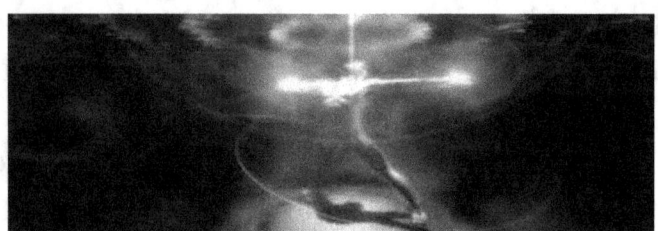

Electron Example

Let me give you a TINY example. According to many Anthropic Scientists, an observer is very deeply involved in any quantum event. On the simplest terms, an electron cloud surrounds a central core of an Atom. Rather than spinning, it is now known <u>to be in superposition</u> as something we could call a wave-packet. It is located <u>at all locations around the nucleus.</u> This only changes when one wishes to sense it. Finding a location of an electron takes 2 things. One is there must be a disruption position in some way and the second may be <u>even more important</u> <u>that a cognizant viewer must be integrated in the</u> <u>observation or the electron will stay in superposition</u>. It is, at the moment of observation, by a conscious mind that the electron "chooses" one of the possible locations to materialize in, collapsing its wave-packet and becoming a particle for the split second it takes to be hit by the other particle. Another way of stating this is that electrons are sometimes a wave and sometimes a particle. Our interaction with the universe converts it. In electro-magnetics finding out that light was sometimes a wave and sometimes a particle has been described for many, many years. It is only now understood that particles act exactly like light.

The logical conclusion of Athropics is that Life can be associated with something that could be considered particle-like [Carnal thought, brain control, and muscle reaction] and sometimes wave-like [having the ability to control our destiny and reality].

The wave-like characteristic of Life and Death are the keys to happiness, understanding, reduction of fear, and oneness with our creator, but for this study, it is what defines light for us.

As we usually are subject to the combined consciousness, or Red is almost always what everyone else sees. The main theme with respect to anthropic science is that something inside entities that are conscious are attached to and help define what we call reality. As someone modifies how he interacts with this group, reality changes. If Carbon 12 was required for the entities, it will be in the reality of those entities. If someone witnesses a tree falling, it will make sound. If no one is around, there is absolutely no need for the tree in the first place so it will not be there. It is sort of written into existence as needed. If we could go to the end of our universe and NOT BE ALIVE, we would not see a star, a galaxy, or even dark matter. Looking back at us there is nothing. I know you are totally skeptical about Anthropics so I thought it would be good to provide some of the MANY issues that have forced this new science into existence. Remember NO ONE wants this to be correct, it just is.

Universe for Light Description

While I have stayed away from the topic of life in most of this book, Anthropics, while mostly dealing with control over our perceived reality, the characterization of Life itself must be examined with both photonic energy and light.

Mass, Lafe, and Photons all are affected the same by time, but they do not directly affect each other. In one way we can say they are all mutually perpendicular and invisible to each other. Only time connects them. The same can be said for Magnetism, Gravity and the living souls. As absolutes of matter, Electromagnetic waves and Life, they are also mutually perpendicular, invisible to each other and are affected by time the same way. [They are not affected.]

The third part of this definition includes Electricity [potential for E-M waves], Aether [Potential for matter], and something we soul call a spirit. [Potential for life] These, certainly, are not affected by time, but are affected by the other two. It seems in our Anthropic world, we have 13 things or dimensions to consider when defining Photons and light. These are:

Life, Soul, and Spirit- *associated with entities that can define, control and augment reality*

Electricity, Magnetism, and E-M Force- *associated with a characterization of In-waves, counteraction of entropy, and the sensation of Light*

Matter, Gravity, and Aether- *associated with the physical construct of our universe.*

The last "dimension" can be called time. *Some just say it is a reference as it has no definition without relativity, but it's an interesting thing just the same.*

Photons and the Soul

In a relative arena, Photons are vibrating electro-magnetics interfaces. As electricity vibrates Magnetic fields are spontaneously produced vibrational stress are exerted from Out-waves from particles when they reach the limit of our universe. Cosmic blasts erupt as out-waves from our universe are converted into in-waves of our linked universe. As out-waves continue to breach the universe, sustainment of electromagnetic energy is assured.

The next question is what makes the vibrational frequency that is needed for our reality. <u>I think you already know the answer and that is the soul collective regulates something we can call universe or reality resonance.</u> Resonance is the frequency of comfort and sustainment of that operational wavelength is established, modified, and continued by that characterization of reality.

Photonic Intrusion into Matter

As we cannot sense photons, our reality adjusts their characterization to make them appear visible at certain times. Without a collective conscious, photons would have no particle component. Keeping with symmetry automatically says that if one measures particles of photons, some equal reduction in matter will be sensed in the world. Luckily, not much matter has to be exchanged for this temporary state to be witnessed.

Photonic Control of Life

We are now learning that photons build life. Optical messaging warns of sickness, harm, and rebuilding of damaged areas and full construction of complex entities. It is this bio-photonic interaction that helps us realize how limited our view of photons has been. Surely photons contribute to light and life, but there still is a question?

What is Light REALLY!!

As I mentioned, the definition of light is not a mass of Photons, in fact many see light in their dreams with their eyes closed. Many see a strong light when they have a near death experience or are in a coma or similar out of reality excursion.

- Light has nothing to do with a certain band of wavelengths that our eyes can make electrical pulses from. As I have been saying, "light" is perceived, not generated. One cannot grasp the fleeting essence of light as it has no reality on its own.

- The importance of Light in ancient texts show us that it in some way helps regulate or even control our perspective of reality.

- To the angels who had lost it, "Light" was <u>more important than life itself.</u>

- If light has a dual in an adjacent universe <u>the dual would be darkness.</u>

- <u>The force of the released soul seems to be associated with light.</u> When people temporarily die they see a massive light.

- The Holy Spirit of God was identified as the light, and the Heaven War rebels were punished by having their "light" removed so they would be stuck on earth.

- Light was the second of 3 creations of God, which includes something we call vibration, and something we call life. Each of the creations seemed to have captured a 3idimensional dynamo of dimensions that control our reality. Together the three make up our world.

- Biblically God created light well before he created the Sunshine and moonshine so original light was self-generated or nuclear.

- God views our universe transversely as he views the beginning and ending of lifetimes instantaneously across lateral time or light. Life and matter cannot breach the speed of light boundary because mass cannot be established nor can life.

Light is as important as Vibrations that make everything and Life itself. We know it's important by the shear inference given in our most holy book.

Light seems to be Electro-magnetic, nuclear, and spiritual. It holds our "reality" together by establishing the vibrational cord from which the dimensions build.

Light is like God's eyes. It brings us into reality, and away from it as needed. It established a link with God and comfort during trouble. How can anyone think Light is simply photons?

Fish and Water

OK! I wasn't able to give you a conclusive answer, but do you know anyone who has tried as hard as I have? I think one of the problems can be understood by fish. *It seems fish cannot understand water! It has no definition for them because it is life. Water carries motion, breath, food, environment, and establishes a link between other fish.* With all that importance, it still has no meaning to them. It just is.

In the Anthropic world light expands beyond simply vibrational nothing into something that helps sustain some portions of life and our perceived reality as well. We may never have a full understanding of what light is, but we are getting closer every day and that understanding is even beginning to keep us healthier. let's revisit Einstein and get a little more insight.

Einstein and Anthropics

As I mentioned at the beginning several things weighed heavy on Einstein in the mid-fifties that made him believe that all of his work was useless. I already explained that Einstein's relativity actually helped define Anthropic control so he was very close. The biggest thing to disturb him was something called quantized Red-shifts.

Quantized Red-shifts

Astronomers have confirmed that galaxy red-shifts are quantized. Here is the seemingly weird part. According to Hubble's law, red-shifts are proportional to the distances. Therefore galaxies must be grouped into spherical shells, which is odd enough. It gets even odder because <u>everything seems to be concentric around our Milky Way galaxy</u>. It's as if the Earth is the center of the Universe. These shells of matter, apparently, are around a million light years apart emanating outward from us. It has been stated that the odds for the Earth having such a unique position in the universe by accident are less than one in a trillion, but it's worse than that.

No Big Bang

Let me tell you what this Earth centered red-shift quantum REALLY means. There could not have been a

BIG Bang. Impossible, impossible! If it had occurred, then there would be nothing where the Earth is today. It would be part of the explosion epicenter. Einstein just could not rationalize the center of the Universe being the Earth. As the great Einstein started doubting everything that he had accomplished, let's see what he had to say.

All these fifty years of conscious brooding have brought me no nearer to the answer to the question, "What are light quanta?" Nowadays every Tom, Dick and Harry thinks he knows it, but he is mistaken. I consider it quite possible that physics cannot be based on the field concept, i.e., on continuous structures. In that case, nothing remains of my entire castle in the air, gravitation theory included, and the rest of modern physics.

I know he sounds like a cry-baby, but he is right. If we want to even attempt to understand what light is, we cannot be thinking that particles exist and we have to come to some resolution about this Earth centered Universe along with all the rest of the things that Einstein finally realized as being impossible.

I apologize about the cry-baby remark. Einstein was a great thinker and he knew when he had a problem. When he found out, he didn't bury his head.

Let me back away from this for a while so no one bashes my head in and go back to our own universe where light can be created by us just being there.

Faith Light

I'm sorry fro this section, but I think you need to understand the bigger picture or you will be stuck with photons bouncing off objects as the definition of light. To really understand light, we have to tackle Anthropic Science because Light really doesn't exist in the same way a wall or tree exists. Anthropic Science says nothing exists; not matter, not energy, not light, without a cognizant observer. This does not mean if a dog sees something it exists. Cognizant, in this sense, means having a connection to the collective "of cognizant souls who regulate reality". Each person has one part of it. One way of thinking would be that the REAL you is your "soul" and the "self" part of you is just a wrapping to enjoy the sensations established by the cognition of souls. Connection to the collective is called faith. It is this faith that makes us cognitive. Most people almost never tap into this light building resource but it is there for us just the same. Faith can be, generally, defined as the "actions associated with Anthropic control of reality". According to Biblical testimony and other religious and historical references; with this Anthropic control, or faith, one can walk on water, turn water into wine, make entire mountains move, pull demons out of people, heal the sick, and produce light or make blind people see light. All these are similar even though they don't sound similar.

According to Hebrews11- **Faith** *means being certain of the things we want to change and acceptance that those things are real even if we do not see it.*

Please understand one thing. The faith associated with Anthropic Control of the environment is not the same thing as "faith in God's Promise", nor was faith in God defined as a prerequisite for walking on water or any of the other faith related things described in the Biblical testimony. If you want to make blind people see light or make your body emit a halo of visible light, these things can be accomplished by removing "self" and freeing your soul up to modify the reality around you. Sorry for sounding so much like someone tripping on LSD or something, but every year we find out more and more about the innate capabilities of the part of us that establishes anthropic control. Some call it self-actualism, others claim "if you think about becoming rich you will become rich". Called the power-of-positive-thinking by others, control of the soul entity can allow people to do and create unimaginable things including light. There are many Gurus that heal sick people by faith and the examples of many different individuals lifting automobiles off an injured people by faith are real. I know you have read stories indicating that adrenaline somehow makes the bones so strong they don't crack in half when lifting 3 thousand pounds, but ---bones will break. Adrenaline is not concrete but faith can be. What faith is, is being attuned with the collective consciousness and changing reality around you. The power-of-positive-thinking is a real thing. If you convince yourself strongly enough, you can affect reality. That is not to say that faith in the creator is not a great thing, but that affects people after death. While we are in this reality, the writers of the Bible and other ancient texts were telling us so much that we disregard. One this is Anthropic control can create or manipulate light. When you get a clear image of a solution to a problem. The creation you see is light. When you see yourself succeeding in some task before it is done. That is light. Everyone does not have to see what you see before it is light. Have you ever gone out in the bright sunlight and everything goes white so you see nothing? That is the absence of light cognition while you are alive, but death may not even halt the production of light.

Dead Control of Light

While I am on this short rant let me talk about those whose "self:" has died. Souls don't die. We are told most of them are sleeping, but demons are just souls that are not asleep and not part of a "self". Yes! They come from the ancient Anak people who died and were punished by God to make their souls wander the earth but the idea is the same cognizant souls can establish light. If a demon possesses someone to experience pleasures of the "Self"; that soul of the dead could, very well, assert some control over our collective reality including making their host unbelievably strong, making them float, and establishing light become real. I know you are thinking some of this was just made up the other day so let's go back in time a little and review documents trying to give us this information thousands of years ago. I'm going to show a few Biblical texts and many other ancient historical references to "<u>Anthropic Light</u>" the ability to sense our true reality.

Anthropic Light

As I mentioned before God made light after he had created mass as was identified in the 1st chapter of Genesis. Only after 4 Ages did he finally make the sun. As I mentioned before, when Moses came down from visiting with God his face glowed as if he WAS light. As you look at some of these discussions that are clearly not talking about reflection of photons, understand that this is something much MORE.

Biblical Light

"Genesis"- *"God said let there be light and there was light"* [This was before the sun light was made.] In the catholic *"Prayer for Enlightenment"* we find *"O Holy Ghost, divine Spirit of light"* [The Holy Ghost was introduced to allow understanding and return the light to humans.] "Ezekiel" 1:4-14 tells us about flying saucers were filled and were powered by light. —*" Out of the midst of fire, of the whirlwind, came four living creatures with four wings. They had straight feet and they sparkled. Their wings were joined together. They did not turn. They all went straight and their wings were stretched upward. Two wings from each were joined together and two covered their bodies. Their appearance was like burning coals of fire. They went up and down* and *out of the fire came lightning. They were as fast as lightning.*

Navaho, Pintes, and Hopi

There traditions tell us tell us about very ancient weapons made of light –*They all told of the Golden Strangers from the sky that came in flying canoes which were armed with Burning Rays of light.*

Iranian Text

This comes from the sacred Zadspram- *"From the seed which was the ox's, they would carry off from it and the brilliance of light was entrusted to the angel of the moon in a place that seed was thoroughly purified by the light and was restored in its many qualities."* [This segment was after the angels had corrupted almost all the animals. In order to reconstruct the animals, God had to put in more "Light".]

Gnostic Texts

This comes from the Gnostic book *"Apocalypses Moses"* Chapter 33: 2- *"And she [Eve] gazed steadfastly into heaven and beheld a chariot of light borne by 4 bright eagles- and angels going before the chariot. In this section* [Eve watched Adam being taken into heaven on a chariot of light after his death].

This verse comes from the *"Book of Abraham"* chapter 4:3-*"And they said, let there be light, and there was light. The gods **comprehended** the light.* [This is an expansion of the Genesis statements. Remember this is before the sun was remade. The odd part is the comprehend word. It seems that it is suggesting light was more than something visible that all could comprehend.]

"Origins of the World"

"Origins of the World" had this to say. - *"The troublemaker that was below them all destroyed the heaven and his Earth. And the six heavens shook violently; for the forces of chaos knew who it was that had destroyed the heaven that was below them. And when Pistis* [one of the trinity] *knew about the breakage resulting from the disturbance, she sent forth her breath and bound him and cast him down into Tartaros* [Hell] *and when they had become disturbed, they made a great war in the seven heavens. Then when Pistis Sophia had seen the war, she dispatched seven archangels to Sabaoth from her light. They snatched him up to the seventh heaven.* [This has the Heaven War, the troublemaker {Satan} and this odd thing called the light.]--- *First she [Eve] was pregnant with Abel, by the first ruler [Adam]. And it was by the angels that she bore the other offspring [Cain]. -the first mother might bear within her every seed, being mixed and being fitted so that the modeled forms might become **enclosures of the light**,"* [This seems to indicate that the losers of the Heaven War and Adamic humans had children together so that the mixed breed might have the "Light".--------
*"And he said, 'Come, let us create a man according to the image of God and according to our likeness, that his image may become **a light** for us.'* [The losers of the Heaven War believed that man would somehow get them

back the "light" that they lost in the war.] *His intelligence was greater than that of those who had made him. And they recognized that he was filled with light* [This was talking about some type of luminous essence in the descendants of Adam.] *The blessed One, sent, a helper to Adam, luminous Epinoia [holy spirit] who is called Life. And she assists by teaching him about the way of ascent. ----But the Epinoia* [Holy Spirit] *of the light which was in him, she is the one who was to awaken his thinking.* [The spirit inside the descendants of Adam was somehow associated with LIGHT.]

Cayce Confirms Genesis

Edger Cayce was a 20th century seer. The interesting thing is that many of his seeings have come true and some were extremely detailed so let's see what he had to say. - *"God moved and said, "Let there be light", and there was light. Not the light of the sun, but rather the light which—through which—in which—every soul had, and has and ever had, its being."*

There can be little confusion with the above statement. This is certainly talking about a linking with souls that we call Anthropics.

"Generations of Adam"

The Essene Jews were meticulous in copying ancient texts over and over. This is one of the most copied texts. "Generations of Adam" had information about LIGHT. [Chapter 8:13-14] *"A **bright light** shown from heaven illuminating the whole palace of king Canaan, and a* *mighty noise from heaven shook the air. Whence the palace stood there was only dust. A war broke out among the people. Armies devastated the land. The people suffered great desolation. There was destruction everywhere.* [Horrible wars and weapons of mass destruction were known and the bright light from heaven that caused the air to shake really showed the power of light, which makes sense if we are talking about modifying REALITY with this light.]

Jubilees

The book of "Jubilees" is still considered canon in some orthodox Bibles of today. This comes from chapter 2:9- *"Nor may we take revenge on him because he has <u>stripped us of the "light"</u>. He marked out the borders of the world and created man in his own image with whom he hopes **<u>again</u>** to people heaven, with pure souls."* [Not only note that without the light, the losers of the Heaven War could not take vengeance on any of the heavenly host, they lost some substantial power without this light thing. Also note that the word "again" is put in the verse to let us know that man was here before Satan's Heaven War and it was recreated after it was over.]

Greek Legend

In discussions about battles between the gods we find the following: *"Hot vapor lapped the titans, flames unspeakable rose bright to the upper air [outer space], **<u>lightning</u>** <u>blinded their eyes</u>."* [Apparently light was and is very powerful and again had nothing to do with the reflections of the Sun's emissions.]

Brazilian Light

In Brazil, we also find a similar story from the Manacitas Tribe. They told and retold ancient stories about flying. One of their cherished legends talks about *the Macunbeiros, which were flying wizards that flew inside circular,* **luminous,** *machines.* [One was to take this is that the wizards could emanate this light or become one with "reality itself.]

Let's See What We Have from Ancient Texts

- Light was created before the sun.

- Light was inside the angels that fought in the Heaven Wars and it was removed as punishment when they lost as if they could no longer affect reality the way they could before.

- Light is somehow associated with an essence inside the descendants of Adam that is associated with transferring to the universe known as heaven or back to our reality.

- The survivors of the Heaven War tried their best to re-associate this Light thing by association with the descendants of Adam.

- The Light described in these verses is much more powerful than what we think of as "normal light".

Creation of Light

I know it's already confusing, but I need to bring all this into the discussions because we need to get rid of the belief that all matter is solid and begin to appreciate this thing called vibration. Let's look back at the Bible for a minute. All of you have read the first chapter of Genesis where the world was made, but I'm thinking you didn't really read it.

Genesis 1:1 The War

Genesis 1:1 *"In the beginning God created the heaven and the earth and the earth became without form and void"* The book of "Jeremiah" tells us that the reason the earth got so bad was that Satan had the war and all the cities in the world were destroyed. Stay with me here. This will make sense in just a little bit.

Jeremiah 4:23-27-[description of the end of the Heaven Wars] *"I beheld the Earth, and, lo, it was without form, and void; and the heavens, and they <u>had no light</u>. I beheld, and, lo, there was no man, and **all the cities thereof were broken down**."*

Genesis 1:2 No Light

Genesis 1:2 *Then <u>darkness was on the face of the deep</u> [no light]and the spirit of Elohiym moved on the face of the waters.*

Genesis 1:3 Light Before Light

1:3- *Then God said <u>let there be light</u>. God divided the light from the darkness.*

Here is the interesting part of this verse. God made light before the Sun, moon and stars were shining in verse 16 much later in time.

Genesis 1:16-19-*And God made two great lights: He made the stars also. And God set them in the firmament of the heaven to give light upon the earth, and the evening and the morning were the fourth day.*

Clearly this light thing described in the 2nd and third verse was not light from the sun, but it was something. I could get into my philosophy about what this first light was, but let me just say for his discussion that, apparently, God created three things

- **Vibration-** to make light and matter
- **Life-** to establish a reality needing light
- **Light-** not photons, but that undefinable characteristic that converted vibration into images, sounds, feeling, understanding, hopes and dreams.

Maybe I still need to broaden your vision of light. For that expansion, let me first restate the Biblical description.

1. The Bible states that when the rebel watchers lost the Heaven War something very special was taken away from the rebels. They were turned into humans as one of the punishments, but the second one seemingly even more severe. The Bile indicates that the **LIGHT** was taken from them.

2. Another place Jesus said he was the **LIGHT** the truth, the way.

3. Jesus also said, *"Believe in the **light** while you have the **light**, so that you may become children of **light**."*

These things sound like Moses and the other writers had no idea what they were writing in Genesis and other parts of our Bible. I believe that God was trying to tell us something about our universe and what LIGHT really was. After all; if you can't trust God, who can you trust?

In all cases we keep coming back to a definition of light that has to do with life.

Light and Life

Hopefully, you can see that, somehow, <u>light and life are connected</u> together just like potential energy and kinetic energy are linked. It would have been nice if Niels Bohr had finished his work on how light and life are interrelated, but he died so you'll simply have to make do with my definition for now.

There is a <u>symmetry of light to be considered</u>. One might say that <u>the outward expression of light causes the inbound condition we might call darkness or "inverse-light"</u>, this secondary light described in the Bible and other ancient texts ties the "inverse-Light" to human life.

You need time to think about this so put the book down and walk around for a moment.

Have you wondered about the similarity of magnetism and gravity? Well, I have, and there is this <u>same similarity between Light and Conscious Life</u> as well. We need to recognize this link as we go further because light references are pretty hard to follow because many things get mottled in figurative speech.

> *I think it is safe to indicate that light has a duality with living.*

We can feel the similarity and certainly the idea of death and darkness having a similarity seems reasonable---just because.

If Gravity [secondary dimension of mass] is associated with Magnetism [secondary dimension of photonic energy] then it should not be a stretch to say that light [a tertiary dimension of photonic Energy] is associated with the Soul [the tertiary dimension of life itself]. As the physical word "light" increases, the intersection of the soul decreases.

> *What this means is that as one focuses on self, he has less influence on reality and his understanding of LIGHT becomes less. Are you one of those that don't understand light? Maybe you really don't understand life.*

With the new revelations and something very strange we call light, let's see if we have any conclusions that could be useful.

Conclusions

The descriptions of what Photons and light are in this book, more than likely, were different than you had initially thought. While many of the descriptions derived here are not so very useful for you when working with light in a normal way, hopefully, you are more aware of its awesome possibilities and how we should not simply turn on a light bulb and not be in awe at its importance. It is my hope that the following were among the bits of information that is carried away and used to expand your awareness.

- If electrons are pushed away from their nucleus, the collapse of the electron becomes light. We call it <u>laser light or the spectral emission of a material.</u>

- Compressing a crystalline lattice produces light and electricity. This is called Piezo-Electric effect or <u>Piezo-Electric Light.</u>

- Electromagnetic waves disrupt and modify our brains when the vibrating frequencies are low One could say they make us <u>Light headed.</u>

- When the frequencies are very high the light can kill a person as <u>Cosmic Light.</u>

- Changing the frequency changes the perceived color, whatever color is. It is called <u>Chromatic Light</u>

- Photons seem to go faster than the speed of light which means they would be going back in time.

- At the speed of light mass goes to infinity, but photons seem to have NO MASS??? It seems so odd that the dimension that makes us see things can become invisible and not seen. [<u>nonexistent Light</u>]

- The atom is not the essence of matter just like a photon is not the <u>essence of light</u>. They are simply EASY models that show some of the characteristics.

- Photons may be nothing more than a vibrating quasi mass. Whatever that is. We could call this <u>Aethereal Light</u>.

- Photonic Force is one of 3 dimensional controls that allow motion in our universe. <u>Dimensional Light.</u>

- All our cells produce and interpret something called <u>Bio-photonic Light.</u>

- If photons have no mass, where does the energy come from? If they do have mass, how in the world can they travel so fast? Anthropics helps us understand.

- <u>Light vibrates faster than the speed of light</u>, but its mass does not go to infinity. While the reasons are not completely understood one thing about light is that it seems to be using backward and forward time together and this allows it to disappear and appear here as needed.

- Lateral time is the ability to view our world at the speed of light. When viewing mass in lateral time, vibrations appear to be halted. Light tracks a timeline that is adjacent to our normal timeline. If we view normal slow moving or stationary mass in lateral time, the mass goes to infinity. If we view mass going the speed of light in normal time, he mass goes to infinity. In Lateral time, the beginning and ending of a lifetime can be viewed simultaneously, in fact, the beginning and ending of the universe can be witnessed.

- Time dilation slows down vibrations and reduces the energy in a dimension. If it is not linked with others, the resonance of the universe [entropy] increases. Energy from an outside force must replace this energy or the universe will slowly run down. What we see today is that the universe seems to be more vibrant each year. Unfortunately, we cannot judge this as we are linked with any time dilation that occurs.

- When light mass disappears, it must reappear in an adjacent universe. This changing may be a doorway to an adjacent universe. Photons in this universe are photinos in an adjacent one.

- If a person goes the speed of light, and his mass becomes spread along the entire distance traveled, he becomes like light.

- Our universe is built around 3 sets of dimensions, matter dimensions, Life dimensions, and photonic dimensions. They are all linked in some way and yet they all are completely separate.

- Light is not photonic energy- One can actually define photons pretty well even if they go faster than the speed of light, but light is a gift.

- Tamashii modeling allowed us to better understand different materials and their innate vibrations.

- John Hutchison has shown that materials can be completely changes by bombardment of vibrations. Things can turn invisible, levitate, and be completely changed.

- God created 3 things. Vibration, Life, and something we call light. Light is connected to everything in the universe. While God's other 2 creations are no slouchers, light is more important than just about anything around us. Don't take it for granted.

- The miracles described in the Biblical history and other ancient texts appear to be more reasonable every year as we understand more about Anthropics.

- Our universe is not controlled by physics. Universal Physics is modified by some collective consciousness. If a tree falls in the woods and no one is there to witness it, it doesn't make a sound, there is no tree, and there is no light.

- Invisibility can be established in matter, life and light. If two identical light beams are emitted but one is out of phase with the first, there is no perceivable light in our reality. If subatomic particles are out of phase, matter ceases to exist.

- There is no light or matter without an observer.

- Red is perceived by all in a collective consciousness controlled reality.

- A majority of the photonic emissions from your cells are in the ultraviolet region, but sometimes there can be visible light emitted as well.

- If we could see ultra-violet and Infrared we would struggle seeing the outline of individuals.

- Light is part of the building blocks of animals, plants and people.

- People and plants both use optical messaging established by Live DNA in our cells. Misinterpretation can cause cancer.

- One reason there are such varied results from Bio-photonic research is something called Anthropics as the will of an individual can completely change the characteristics of bio-photonic actions.

- Dead and alive DNA are exactly the same except for one critical element. Dead DNA do not emit photons.

I know this book was hard to read and understand, but it's difficult enough to address Photons as finite entities in any responsible way. When I took on the task of describing light, it was even more difficult as no definition can be complete in our understanding. If you want to establish the relative position description, all that is required is to establish different universes for each observing entity and link like velocity observers together in some sort of quantum latch. Who they each perceive the same light might be interesting, but it seems too complex for me. I tried to keep it simple so it would be enjoyable.

The new discoveries of Anthropics and Biophotonics will soon allow us to understand light.

Thanks for reading.

The End

About the Author

Steve Preston is a long lime author of scientific, esoteric facts. His books focus on the painful truths rather than whitewashed details that make us comfortable. If you are interested in the truth instead of comfort, please review other works by Mr. Preston as shown in the following list. The images are some from Egypt taking the older version of taxi. To the right the writer is shown in the Jewish Negev desert of Israel where the Dead Sea Scrolls were found. I made these travels so you wouldn't have to.

To the left below are a couple of pictures as we searched the New Zealand caves possibly visited by the ancient Maori and the last image is of the author investigating the statues on the Acropolis in Athens Greece. Luckily light was everywhere I looked.

History of Mankind Series
20th Century To The End Of Time
The Second Creation of Man
The Creation of Adam and Eve
A New View of Modern History
Close Look at Ancient History
The Antediluvian War Years
The First Creation of Man
Man After The Flood

Modern American Topics
History of Powerful Women
Promote the General Welfare
Modern Misconceptions
Our Very Odd Presidents
American School Disaster
The Bad Side of Lincoln
Can We Save America?
Great American Quiz
Humans on Display
Consensus Science
Monsters are Alive
US History Errors

Prehistoric America
Who Discovered the Americas?
Mysterious PreIncan Journey
Phoenicia and the Lost Jews
Romans found America

Prehistoric Technology
Amazing Technology
Mysterious Pyramids
Incredible Titans
Anakim Gods

Prehistoric History
Creation and Death of Dinosaurs
Kingdoms Before the Flood
When Giants Ruled the Earth
Not from Space

Reality Science Anomaly
Our 12-Dimensional Universe
Mystery of Photons and Light
Meaning of Life and Light
Incredible Nikola Tesla
Is Time Travel Possible?
Biophotonics and Healing
Vibrational Matter
Slip Through a Wall
Anthropic Reality

Historical Fiction
Conrad and the Flood
Secrets of Washington
Shama and the Tower
Naille and the Exodus
Religious Anomalies

Biblical History
Does Science Confirm the Bible?
History Confirmed By The Bible
Abraham to Moses
Adam to Abraham
Adam's First Wife
Moses to Jesus

Moses Studies
Moses Story Part 1
Moses Story Part 2
Expanded Genesis
Exploring Exodus
Exploring Genesis

Christian Studies
Differences in the King James Bible
Why the King James Bible Failed
Understand the New Testament
Old Testament Used By Jesus
New Testament Mysteries
Allah' God of the Moon
Errors in Understanding
Old Testament Mysteries
New look at the Bible
Incarnations of God

Biologic Anomaly
Tracing Cro-Magnon to Jesus
God Didn't Make The Ape
DNA of Our Ancestors
Homo Erectus as a Man
DNA Anomalies
Races of Men
Lizard People

Wars
America's Civil War Lie
Behind the Tower of Babel
World War with Heaven
Four Armageddons
World War Before
World War Zero
Six Deaths of Man
Driven Underground

Egyptian Studies
Truth About Hyksos Pharaohs
Scythians Conquered Ireland
Mysteries of the Exodus
Egyptian Foreigners
Moses Saved Egypt
Secrets of Thoth

Metaphysic Science Anomalies
Releasing Your Consciousness
Understand your Heart
Vampires among Us

Awaken the Departed
Self, Soul, Spirit
Life Resonance
Self-Virtualization
Of Heaven and Hell
True Happiness

Flight & Space Travel
Ancient History of Flying
Anomalies in Flight
Living on Venus
Space Anomalies
Where UFOs Go
Martians

Angels and Demons
Sex Crazed Angels
The Antichrist
The Devil

www.ingramcontent.com/pod-product-compliance
Lightning Source LLC
Chambersburg PA
CBHW072304200526
45168CB00014B/350